Basic Machinery Vibrations

An Introduction to Machine Testing, Analysis, and Monitoring

Author: Ronald L. Eshleman, Ph.D., P.E.
Director, Vibration Institute

Editor: Judith Nagle-Eshleman, Ph.D.
Secretary-Treasurer, Vibration Institute

VIPress, Incorporated
Clarendon Hills, Illinois 60514

Eshleman, Ronald L.
　　Basic Machinery Vibrations: An Introduction to Machine Testing, Analysis, and Monitoring/Ronald L. Eshleman
　　p.　cm　VIPress, Inc.
　　Includes bibliographical references and index
　　ISBN 0-9669500-0-3

　　1. Machinery---Monitoring　2. Machinery---Analysis
　　I. Title

© 1999 VIPress, Incorporated, Clarendon Hills, IL 60514

All rights reserved. No part of this book may be reproduced in any form or any media without the written permission of the publisher.

The author and publisher have done their best in preparing this book. Their efforts include the development and testing of theories and data contained herein. The author and publisher make no warranty, expressed or implied, with regard to the methods and data contained in this book. The author and publisher are not liable in the event of incidental or consequential damages in connection with the application of the technology contained herein.

All product names mentioned herein are the trademarks of their respective owners.

Printed in the United States of America

10 9 8 7 6 5 4 3 2 1

TABLE of CONTENTS

Chapter I: Basic Machinery Vibrations

The Physical Nature of Vibration	1.1
Vibratory Motion	1.2
Measures	1.6
Vibration Measurement	1.10
Phase Measurement	1.12
Vibration Analysis	1.13
Excitation	1.14
Natural Frequencies, Mode Shapes, and Critical Speeds	1.15
Summary of Basic Vibrations	1.16

Chapter II: Data Acquisition

Selecting a Measure	2.1
Vibration Transducers	2.4
Triggering Devices	2.8
Transducer Selection	2.10
Transducer Mounting	2.11
Transducer Location	2.11
Frequency Spans	2.13
Data Display	2.13
Summary of Data Acquisition	2.15
References	2.16

Chapter III: Data Processing

Oscilloscopes	3.1
FFT Analyzer	3.3
Electronic Data Collector	3.5
Data Sampling	3.5
Aliasing	3.7
Windowing	3.8
Dynamic Range	3.11
Averaging	3.12
Setup of FFT Analyzer and Data Collector	3.14
Summary of Data Processing	3.17
Reference	3.18

Chapter IV: Fault Diagnosis

Fault Diagnosis Techniques	4.1
Operating Speed Faults	4.6
Rolling Element Bearings	4.14
Gearboxes	4.20
Electric Motors	4.24
Centrifugal and Axial Machines	4.30
Pumps	4.31
Fans	4.36
Compressors	4.39
Summary of Fault Diagnosis	4.39
References	4.41

Chapter V: Machine Condition Evaluation

Shaft Vibration	5.2
Bearing Vibration	5.3
Casing Vibration	5.5
Summary of Machine Condition Evaluation	5.8
References	5.9

Chapter VI: Machine Testing

Test Plans	6.1
Selection of Test Equipment	6.3
Site Inspection	6.4
Acceptance Tests	6.4
Baseline Tests	6.4
Resonance and Critical Speed Testing	6.5
Fault, Condition, and Balance Tests	6.11
Specifications	6.11
Environment and Mounting	6.12
Presentation of Data	6.12
Reports	6.14
Summary of Machine Testing	6.16
References	6.17

Chapter VII: Periodic Monitoring

Listing and Categorization	7.2
Machinery Knowledge	7.2
Route Selection and Definition	7.5
Measures and Measurement Points	7.8
Baseline Data	7.11
Frequency of Data Collection	7.12

Selection of Test Equipment	7.12
Screening	7.13
Trending	7.15
Alarms	7.16
Reports	7.17
Summary of Periodic Monitoring	7.17
References	7.18

Chapter VIII: Basic Balancing of Rotating Machinery

Types of Unbalance	8.2
Balancing Equipment	8.3
Prebalancing Checks	8.4
Measurements	8.4
Relationship Between Mass Unbalance and Phase	8.6
Trial Weight Selection	8.7
Balancing Pitfalls	8.7
Vector Method with Trial Weight	8.7
Weight Splitting and Consolidation	8.9
Acceptable Vibration Levels	8.9
Summary of Basic Balancing of Rotating Machinery	8.10
References	8.12

PREFACE

Basic Machinery Vibrations is for beginners in the field of predictive maintenance who want to learn the basics of machinery vibrations. The book has been developed to serve as notes for a four-day course in machinery vibrations. Additional examples and review questions may be added at the discretion of the instructor. The summaries at the end of each chapter may also be used for one-and two-day review classes.

The author acknowledges the contributions of data from Kevin R. Guy, David B. Szrom, and Nelson L. Baxter. The proofreading efforts of Loretta G. Twohig and Dave Butchy are appreciated.

Clarendon Hills, Illinois   Ronald L. Eshleman
May, 1999

CHAPTER I
BASIC MACHINERY VIBRATIONS

An analyst without a knowledge of the basics is like a machine with an inferior foundation.

Vibration has traditionally been associated with trouble in machines — wear, malfunction, noise, and structural damage. In more recent years, however, vibration has been used to save industry millions of dollars in machine downtime. Evaluation of changes in levels of machine vibration has become an important part of most maintenance programs. Similar evaluations are used to solve design problems as well as to establish the cause of chronic malfunctions and failures.

The fundamentals of machine vibrations and their measurement are presented in this chapter. Units and terminology are defined. Conversions of amplitude and frequency measures are enumerated. The phase angle between measurement points and its significance are given. Certain machine properties are described.

VIBRATION UNITS

The basic units used in this book to describe vibratory forces and motions are pound *(lb)*, inch *(in.)*, and second *(sec)*.

Amplitudes of vibrating motion are described using the following units:

  displacement, mils-peak to peak (1,000 mils = 1 inch)

  velocity, in./sec-peak or rms (IPS-peak or rms)

  acceleration, g's-peak or rms (386.1 in./sec^2 = 1 g)

Frequencies are expressed in cycles/minute (CPM) or cycles/second (Hz).
Phase is expressed in degrees (deg), in which one revolution of a shaft or one period of vibration is 360°.
Speeds are expressed in revolutions/minute (RPM).

The Physical Nature of Vibration

Machines and structures vibrate in response to one or more pulsating forces, often called excitation. Examples include mass unbalance and misalignment. The process is one of cause and effect (Figure 1.1). The magnitude of vibration is dependent not only on the force but also on properties of the system, both of which may depend on speed. These are mass (weight divided by the gravitational constant), stiffness, and damping; see Figure 1.2. Stiffness depends on the elasticity of the

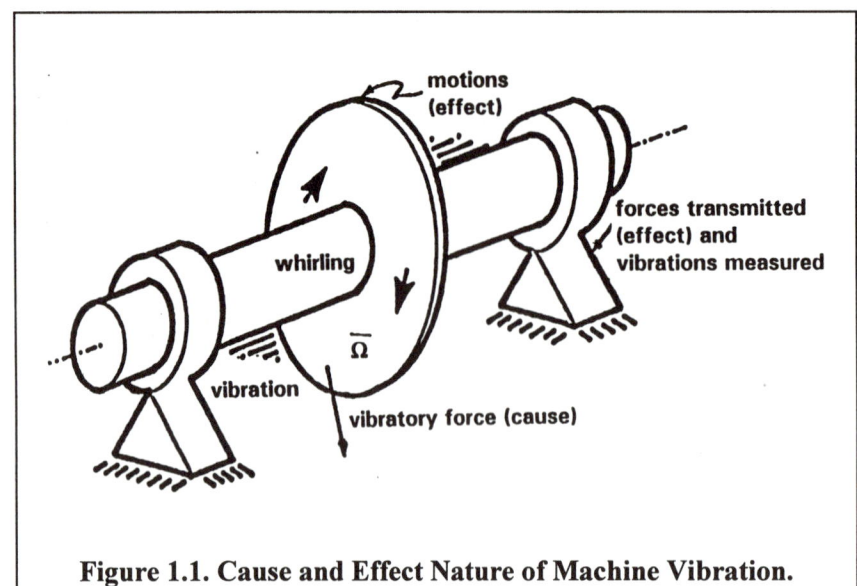

Figure 1.1. Cause and Effect Nature of Machine Vibration.

materials comprising the system and is expressed as force per unit deflection (pounds/inch, lb/in.). Stiffness is determined by placing a known load (in pounds) on a structure while measuring its deflection. Damping is a measure of the ability of a system to dissipate vibration energy. Damping is proportional either to displacement, in the case of structures, or to velocity, for example, shock absorbers and fluid-film bearings. The cause of vibration is usually governed by several factors: the process the machine is designed to perform, manufacturing and installation tolerances, and defects in machine components due to manufacturing and wear. Vibration can be used to identify defects that arise from defective design, faulty installation, and wear.

Vibratory Motion

Three fundamental characteristics of vibration are frequency, amplitude, and phase. *Frequency* is defined as the number of cycles or events per unit time. It is expressed as cycles per second (Hertz, or Hz), cycles per minute (CPM), or orders of operating speed if the vibration is induced by a force at rotational speed. The operating speed of a machine, as well as critical speeds, are expressed in revolutions per minute (RPM). The *period* is obtained from the time waveform (Figure 1.3); it is the reciprocal of frequency. The period is defined as the time required to complete one cycle of vibration. *Amplitude* is the maximum value of vibration at a given location on the machine. *Phase* is the time relationship, measured in degrees, between vibrations of the same frequency (Figure 1.4). From Figure 1.4 it can be seen that the peak vibration measured at point B occurs in time before the peak vibration measured at point A. The vibration at point B is said to lead the vibration at point A. Phase can be used to determine the time relationship between an excitation (force) and the vibration it causes; for example, the force due to mass unbalance and the vibration it causes. This phase is used in balancing.

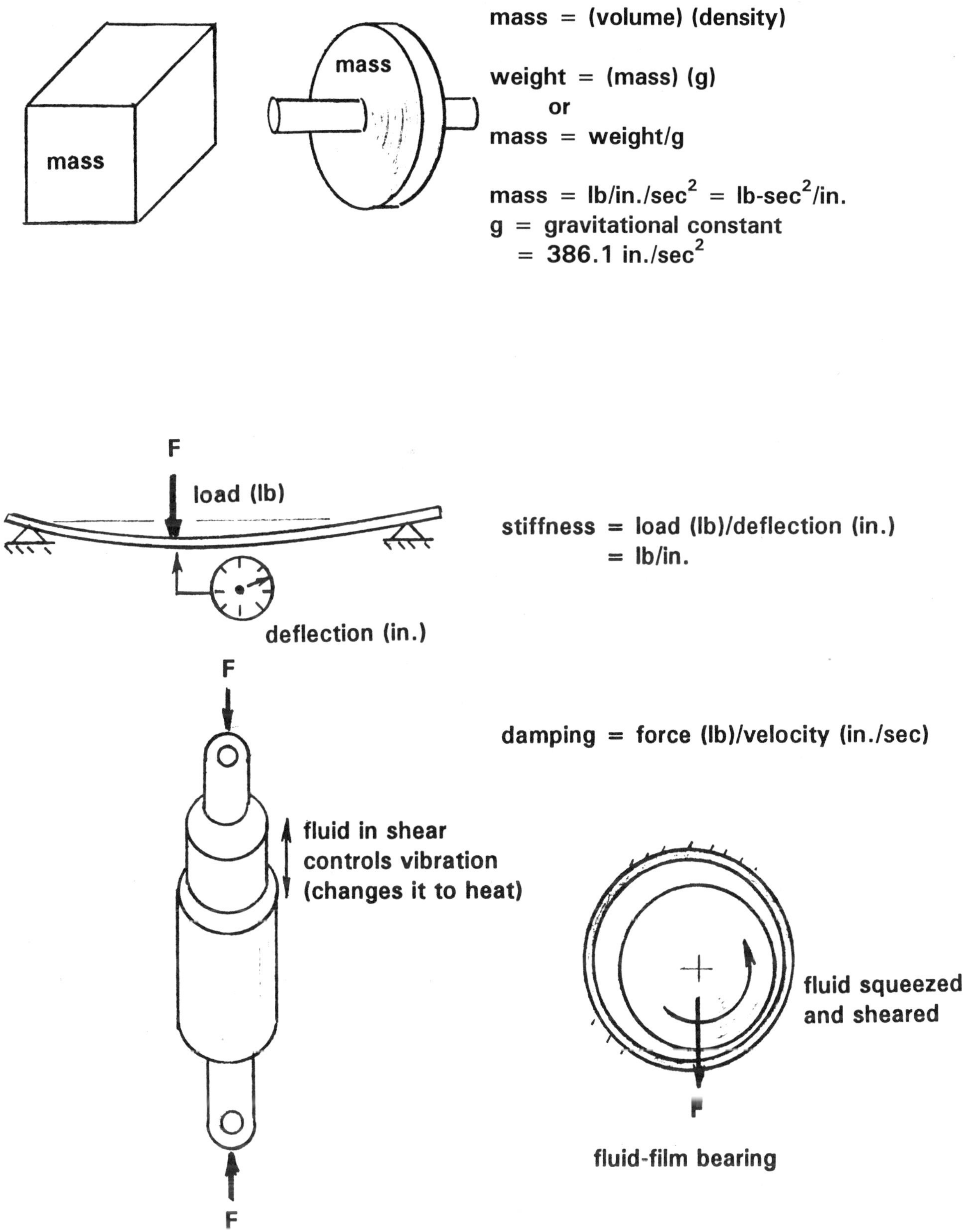

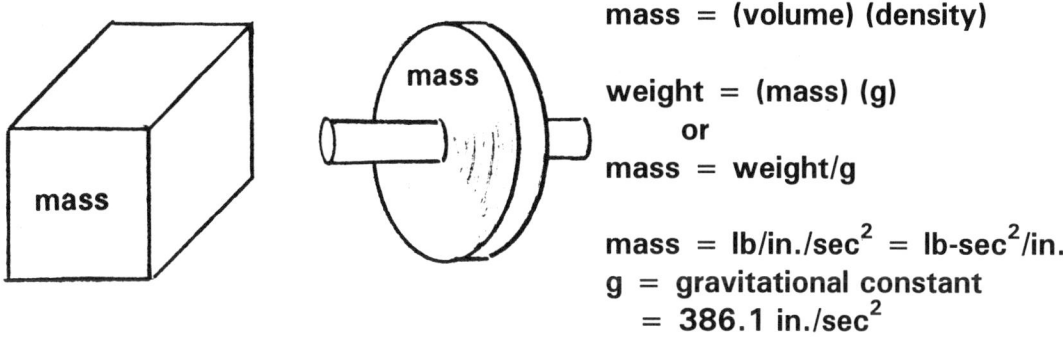

mass
mass
mass = (volume) (density)
weight = (mass) (g)
or
mass = weight/g
mass = lb/in./sec^2 = lb-sec^2/in.
g = gravitational constant
= 386.1 in./sec^2
F
load (lb)
deflection (in.)
stiffness = load (lb)/deflection (in.)
= lb/in.
F
damping = force (lb)/velocity (in./sec)
fluid in shear
controls vibration
(changes it to heat)
fluid squeezed
and sheared
F
fluid-film bearing
F

Figure 1.2. System Properties.

.001 = 1 MILLI SECOND

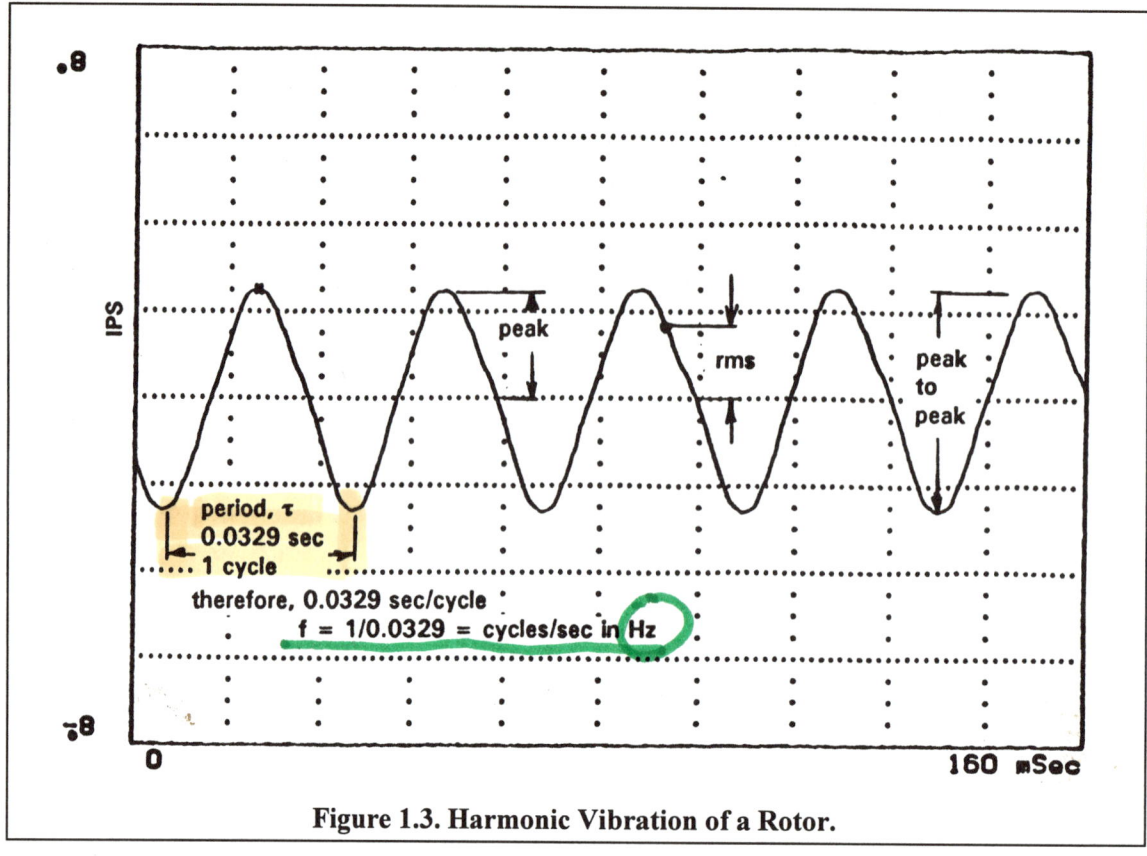

Figure 1.3. Harmonic Vibration of a Rotor.

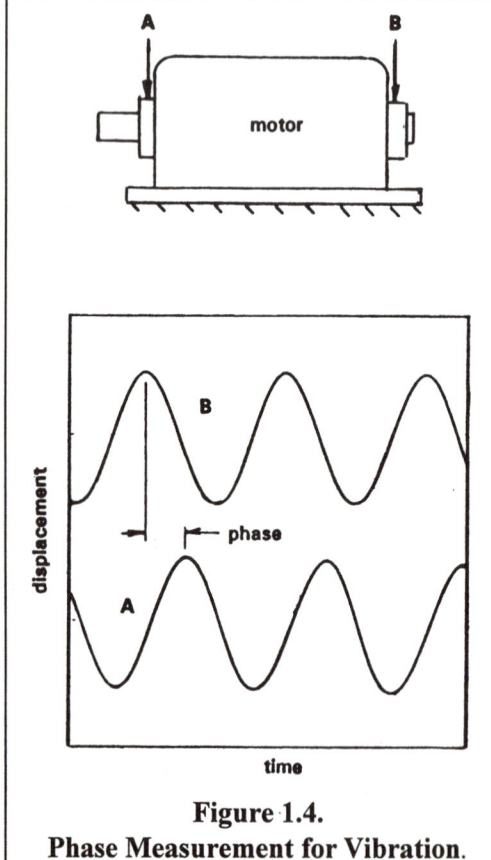

Figure 1.4.
Phase Measurement for Vibration.

Motion that repeats at equal time intervals is said to be periodic (see Figure 1.3). The sine waveform of Figure 1.3 has a period τ. Period is measured in seconds or milliseconds [1,000 milliseconds (msec) are equal to one second (sec); to obtain seconds from milliseconds, move the decimal point to the left three places or divide by 1,000]. The frequency f is equal to the reciprocal of the period, or $1/\tau$. The most basic form of periodic motion is sinusoidal motion (often called harmonic motion), which is represented by a single sine waveform (Figure 1.3). Some vibratory motions of machines are harmonic; an example is the vibration of a machine due to mass unbalance, which occurs at a frequency of operating speed. However, most machines have multiple frequency components in their complex vibration pattern that result in a nonharmonic but periodic waveform, such as that shown in Figure 1.5. Harmonics are integer multiples (i.e., 1, 2, 3, 4, etc.) of any sinusoidal vibration. Orders are integer multiples of sinusoidal vibration at a frequency of the operating speed of the machine.

The amplitude of vibration can be expressed in several ways: as root mean square (rms), peak (p), and peak to peak (p-p); see Figure 1.3 and Figure 1.5. Peak-to-peak amplitude is measured on the time waveform from adjacent positive and negative peaks. For a simple harmonic waveform such as that shown in Figure 1.3, these values can be expressed in terms of rms or peak; rms is

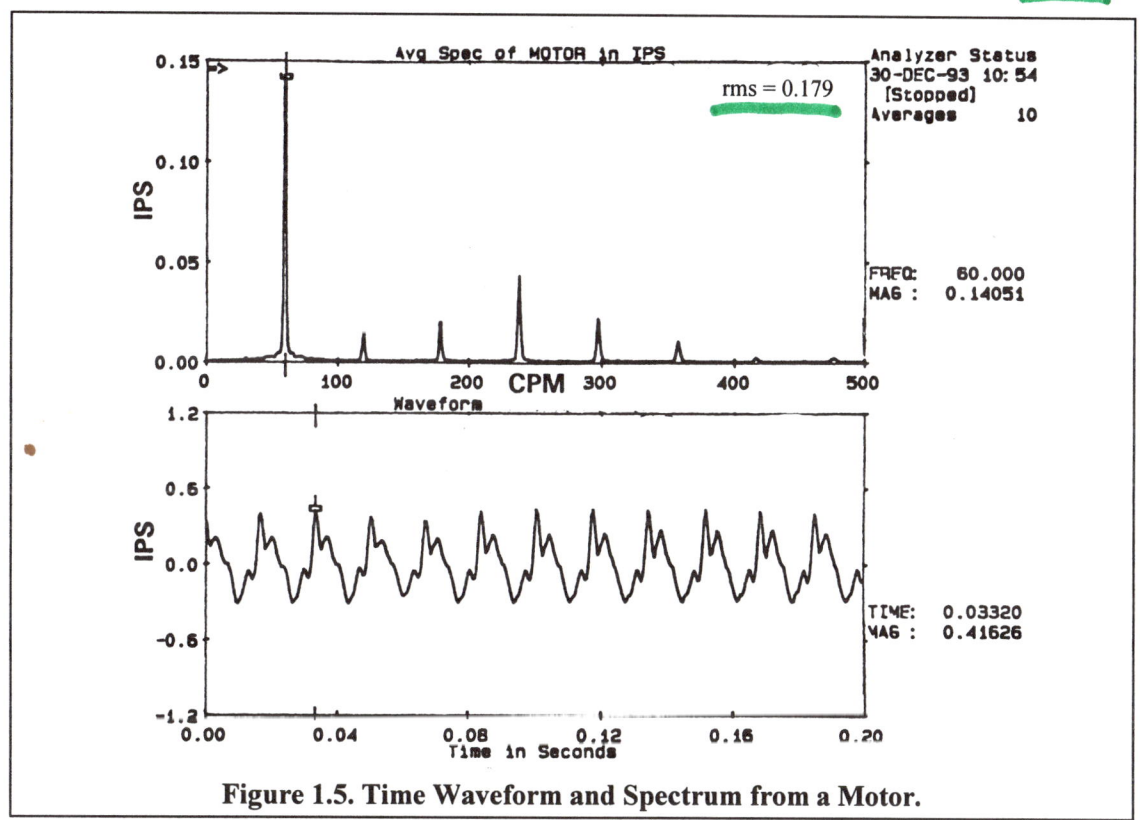

Figure 1.5. Time Waveform and Spectrum from a Motor.

equal to 0.707 peak and peak to peak is two times peak. For any nonharmonic waveform such as that shown in Figure 1.5 rms *cannot* be converted to peak and vice versa. The positive amplitude is not usually equal to the negative amplitude for a nonharmonic waveform. The peak value is the largest positive or negative value. Therefore, the peak-to-peak value will not be equal to two times the peak. Multiplication of rms by 1.414 is not a true peak unless the vibration is harmonic; that is, the vibration contains only one frequency. Many instruments display a peak value that is 1.414 times rms. This is not a true peak value unless the waveform is sinusoidal and contains one frequency. Note that rms relates to the energy of a machine vibration.[1] For example, the rms value of

---

[1] The rms can be defined mathematically as rms = $0.707 \sqrt{(v_1^2 + v_2^2 + v_3^2 + v_4^2 + v_n^2)}$, where v_i are harmonic peak components of vibration and n is the number of components. The rms can also be obtained from special analog electrical circuits.

the waveform shown in Figure 1.5 is 0.179 IPS and the peak value is 0.416 IPS. Note that by multiplying the rms by 1.414 a new value of 0.263 IPS is obtained for the peak. This is often termed the derived peak.

The number of cycles completed in a unit of time is the frequency of the vibration (the reciprocal of the period).

$$\tau = \text{period, seconds/cycle}$$
$$f = 1/\tau, \text{ cycles/second (CPS)}$$
$$N = 60 f, \text{ cycles/minute (CPM)}$$

Vibration with a period of 11.899 msec (0.0119 sec) can be converted to a frequency (84.04 Hz or 5,042 CPM) using the simple equation for frequency $f = 1/\tau$.

Measures

The measures used to evaluate the magnitude or amount of machine vibrations are shown in Table 1.1.

Table 1.1. Vibration Measures.

Measure	Units	Description
displacement	mils p-p*	motion of machine, structure, or rotor, relates to stress
velocity	in./sec	time rate of motion, relates to component fatigue
acceleration	g's**	relates to forces present in components

* 1 mil = 0.001 inch; p-p = peak to peak
** 1 g = 386.1 inches/sec^2

Displacement. Displacement is the dominant measure at low frequencies and is related to stress in flexing members. It is expressed in mils peak to peak because the machine motions are often nonharmonic and therefore yield different positive and negative peaks. Displacement is used as the measure for low-frequency vibration [less than 1,200 CPM (20 Hz)] on bearing caps and structures. Displacement is also commonly used to determine the relative motion between a bearing and its journal or between the machine casing and its shaft. In this case it is used at the frequency of operating speeds and orders. Figure 1.6 shows harmonic displacement and acceleration plotted for a vibration velocity of 0.2 IPS at various frequencies. The displacement of a 0.2 in./sec velocity at 600 CPM (10 Hz) is 6.4 mils peak to peak. The displacement at 60,000 CPM (1,000 Hz) is only 0.064 mil peak to peak. It is therefore difficult to measure displacement at high frequencies because of the small amplitude of the vibration relative to any noise in the signal.

Velocity. Velocity is the time rate of change of displacement. It is dependent upon both displacement and frequency and is related to fatigue. The greater the displacement and/or frequency of vibration, the greater is the severity of machine vibration at the measured location. Velocity is used to evaluate machine condition in the frequency range from 600 CPM (10 Hz) to 60,000 CPM (1,000 Hz).

Acceleration. Acceleration is the dominant measure at higher frequencies. It is proportional to the force on a machine component such as a gear and is used to evaluate machine condition when frequencies exceed 1,000 Hz (60,000 CPM). In Figure 1.6 a 0.2 in./sec vibration at 1,000 Hz is equal to 3.25 g's acceleration. But the acceleration for a 0.2 in./sec vibration at 10 Hz (600 CPM) is only 0.03 g. Acceleration is a poor measure at low frequencies because the signal strength is low.

Conversion between measures. A graphic illustration of the relationship between harmonic displacement, velocity, and acceleration is given in Figure 1.7. For *harmonic motion* the peak values for displacement, velocity, and acceleration can be calculated from the relationships shown at the top of the next page.

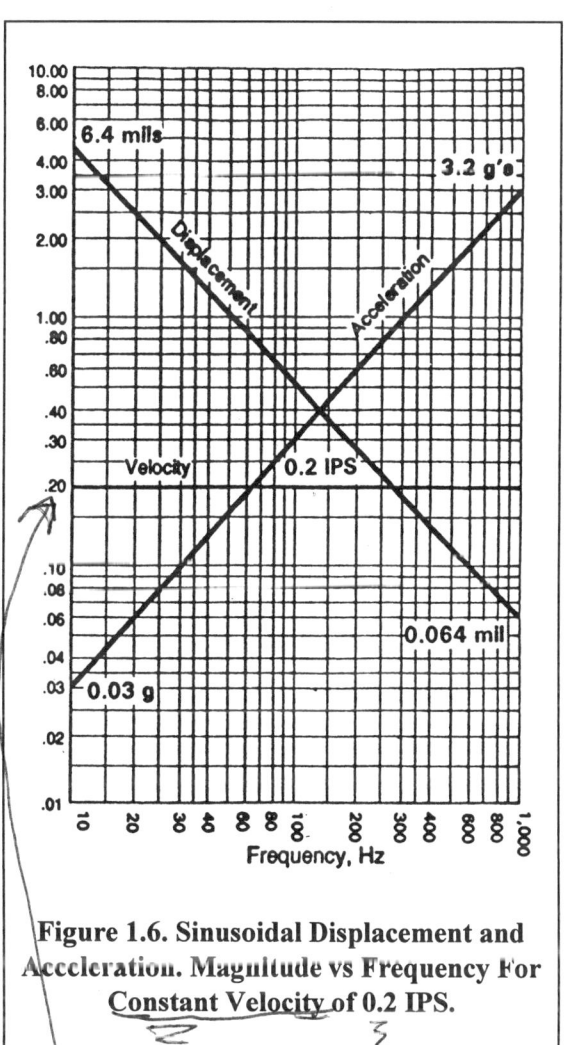

Figure 1.6. Sinusoidal Displacement and Acceleration. Magnitude vs Frequency For Constant Velocity of 0.2 IPS.

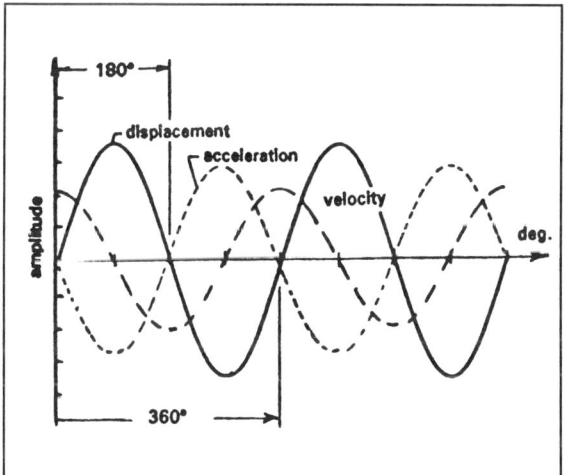

Figure 1.7. The Relationship of Displacement, Velocity, and Acceleration of Motion.

$$\text{Velocity} = 2 \pi f D$$

$$\text{Acceleration} = 2 \pi f V \text{ or } (2 \pi f)^2 D$$

D = peak displacement, inches

f = frequency, cycles/sec

V = velocity, in./sec

A = acceleration, in./sec^2 (divide by 386.1 in./sec^2/g to obtain acceleration in g's)

Note that the phase angle is equal to 90° between the displacement and velocity. The velocity leads the displacement; that is, velocity is ahead of displacement in time (Figure 1.7). The displacement lags the acceleration by 180°. This means that its peak occurs at a later time.

Example 1.1: Convert a displacement of 2 mils peak to peak at 1,775 CPM to velocity in in./sec-peak.

2 mils peak to peak = 1 mil peak = 0.001 inch peak

f = 1,775 CPM; f = 1,775 CPM/60 = 29.58 Hz

Then V = 2 π f D

V = (6.28) (29.58) (0.001) = 0.186 in./sec

Example 1.2: Convert a velocity of 0.15 IPS-peak at 6,000 Hz to acceleration in g's rms.

A = 2 π f V

A = (6.28) (6,000 Hz) (0.15 IPS-peak)

A = 5,652 in./sec^2

A = 5,652/386.1 = 14.64 g's peak

A = (14.64) (0.707) = 10.35 g's rms

In order to convert acceleration to velocity or velocity to displacement the conversion formulas must be rearranged. To convert acceleration to velocity the formula

$$A = 2 \pi f V$$

can be rearranged by dividing both sides of the equation by $2\pi f$ and canceling terms:

$$A/2\pi f = 2\pi f V/2\pi f$$
$$V = A/2\pi f$$

Any of the conversion formulas can be rearranged in this way.

The 386.1 in./sec^2 in Example 1.3 converts g's into in./sec^2. The 2 in the equation changes the peak value into peak to peak. Note that the basic units of the equation are peak inches and seconds.

Example 1.3: Convert an acceleration of 0.5 g peak at 1,775 CPM to displacement in mils peak to peak.

$$f = 1{,}775 \text{ cycles/min } (1 \text{ min}/60 \text{ sec}) = 29.58 \text{ cycles/sec}$$
$$D = \text{acceleration}/(2\pi f)^2$$
$$D - 0.5 \text{ g } (386.1 \text{ in./sec}^2/g)/[(2\pi\, 29.58)^2]$$
$$D = 0.0056\ (2)\ (1{,}000 \text{ mils/in.})$$
$$D = 11.2 \text{ mils peak to peak}$$

Example 1.4: Convert a vibration acceleration of 2 g's rms at 60,000 CPM (1,000 Hz) to vibration velocity in in./sec peak.

$$A = 2\ (1.414) = 2.828 \text{ g's peak}$$
$$A = 2.828 \text{ g's peak } (386.1 \text{ in./sec}^2) = 1{,}091.89 \text{ in./sec}^2 \text{ peak}$$
$$V = 1{,}091.89 \text{ in./sec}^2 \text{ peak}/[2\pi(1{,}000)]$$
$$V = 0.17 \text{ in./sec peak}$$

Example 1.5: Convert a vibration velocity of 0.2 in./sec rms at 120,000 CPM (120 kCPM) to acceleration in g's peak.

$$f = 120{,}000 \text{ CPM}/60 = 2{,}000 \text{ Hz}$$
$$V = 0.2 \text{ in./sec rms } (1.414) = 0.282 \text{ in./sec peak}$$
$$A = 2\pi\ (2{,}000 \text{ Hz})\ (0.282 \text{ in./sec peak})/386.1 \text{ in./sec}^2/g$$
$$A = 9.18 \text{ g's}$$

Vibration Measurement

Mechanical vibration is measured by a transducer (also called a pickup or sensor) that converts vibratory motion to an electrical signal. The units of the electrical signal are volts (v) or, more typically, millivolts (mv). There are 1,000 mv/v; to obtain volts from millivolts move the decimal three places to the left or divide by 1,000. The measured signal in volts is sent to a meter, oscilloscope, or analyzer. Amplitude is calculated by dividing the magnitude of the voltage by a scale factor in mv/mil, mv/IPS, mv/g, mv/deg, or some other ratio of mv/engineering unit that relates to the transducer used. Figure 1.8 is a schematic representation of the common types of transducers available for vibration measurement on a rotor/bearing system.

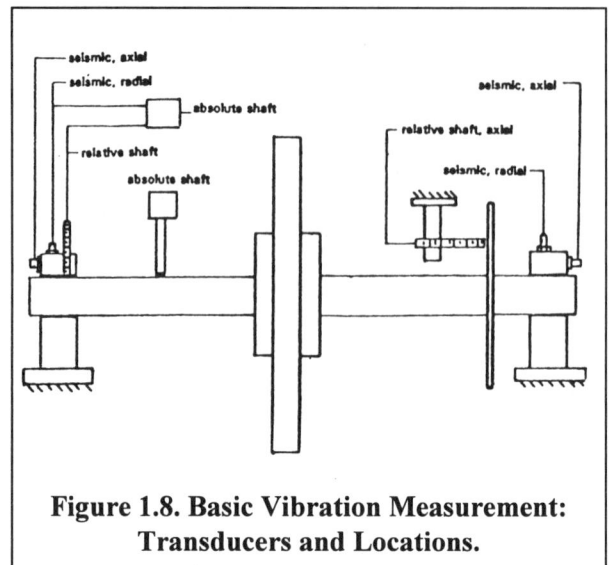

Figure 1.8. Basic Vibration Measurement: Transducers and Locations.

Proximity probes, also termed noncontacting eddy current displacement transducers, are attached to the bearing housing and measure shaft vibration relative to the mounting position of the probe. Two probes are usually mounted at a 90° angle to each other (Figure 1.9). The horizontal probe is always 90° to the right of the vertical probe when viewed from the drive end of the machine. Note that the horizontal vibration leads the vertical by 90° for counterclockwise shaft rotation (Figure 1.9). A *shaft rider* is a device that rides on the shaft and measures its absolute vibration. *Velocity transducers* measure absolute vibration on the bearing cap. The velocity can be integrated to displacement electronically by an integrator or mathematically in the FFT spectrum analyzer. *Accelerometers* measure absolute vibration in g's. This vibration can be integrated to velocity or displacement; however, noise is often a problem at low frequencies in this process. The vibration on a shaft cannot be obtained directly from an absolute reading (i.e., using an accelerometer or a velocity transducer) because of the dynamics of the bearing located between the shaft and the bearing housing. A proximity probe or shaft rider must be used to obtain shaft vibration.

An example of a vibration measurement taken from a vertical water pump is shown in Figure 1.10. The time waveform in millivolts (mv) is taken directly from the vibration transducer. A velocity transducer with a scale factor of 1,000 mv/IPS was used to obtain the waveform. The peak value measured was 934 mv. The peak velocity is therefore 934 mv/(1,000 mv/IPS), or 0.94 IPS.

Two transducers can be used to determine the phase between two locations on a machine, but the location of each transducer must be considered when data are evaluated. The axial transducers

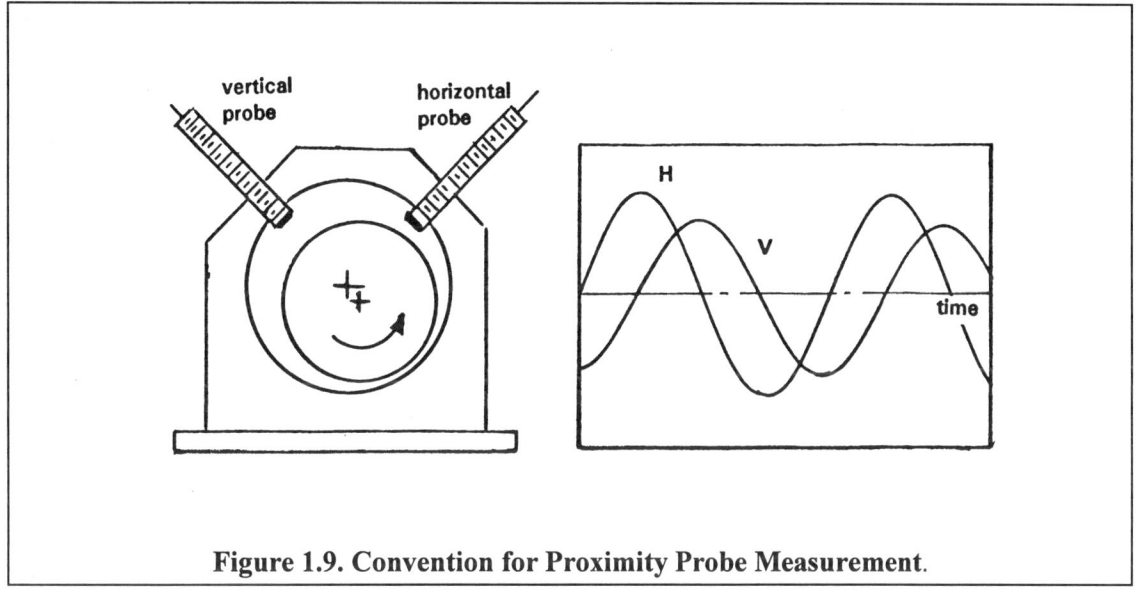

Figure 1.9. Convention for Proximity Probe Measurement.

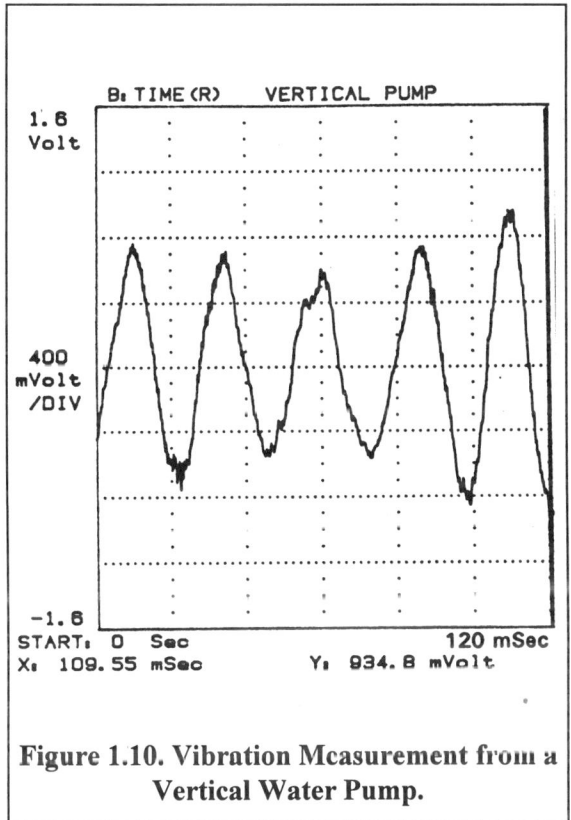

Figure 1.10. Vibration Measurement from a Vertical Water Pump.

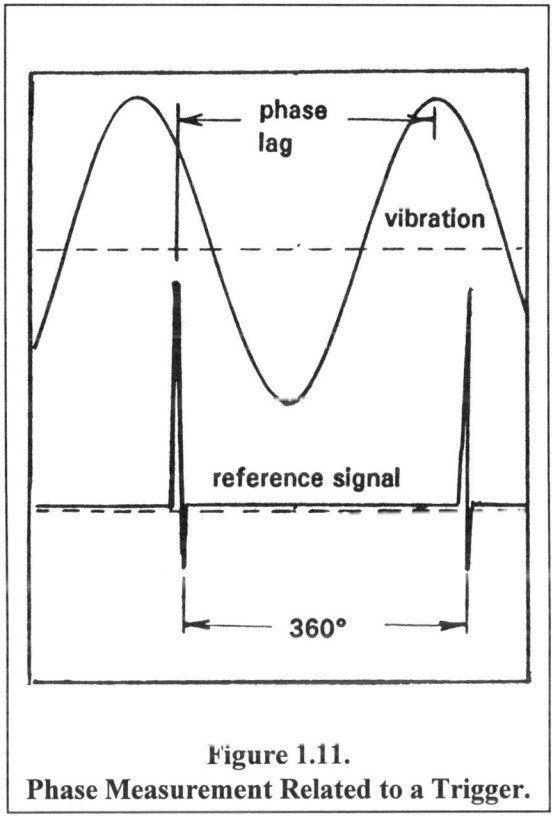

Figure 1.11. Phase Measurement Related to a Trigger.

shown in Figure 1.8 are mounted 180° out of phase. Therefore, 180° must be added to the reading of one transducer.

Phase Measurement

The phase angle between two signals indicates their relationship to each other in time. The two signals can represent either vibration or forces, and their relationship can indicate a condition such as misalignment, the frequency of a critical speed, or the location of the heavy spot on a rotor during balancing.

Phase is measured in the time waveform (amplitude vs time) using an analog or a digital oscilloscope (see Figure 1.4), dual-channel analyzer, phase meter, or strobe light. It is essential to obtain accurate differential time measurements of the signals when measuring phase. Phase is sometimes measured from a reference signal generated once per revolution by a stationary sensor — e.g., an optical pickup, proximity probe (Figure 1.11), or magnetic pickup — looking at optical tape or a keyway on the shaft. The reference signal relates to a unique angular position on the shaft. The phase of the vibration signal can be measured with respect to the angular position on the shaft. The phase angle, which is related to the time required to make one revolution of the shaft, is obtained by multiplying 360° by the time between the two events (reference signal and peak vibration signal) divided by the period of the vibration. This phase angle is measured automatically by the balance analyzers used in balancing.

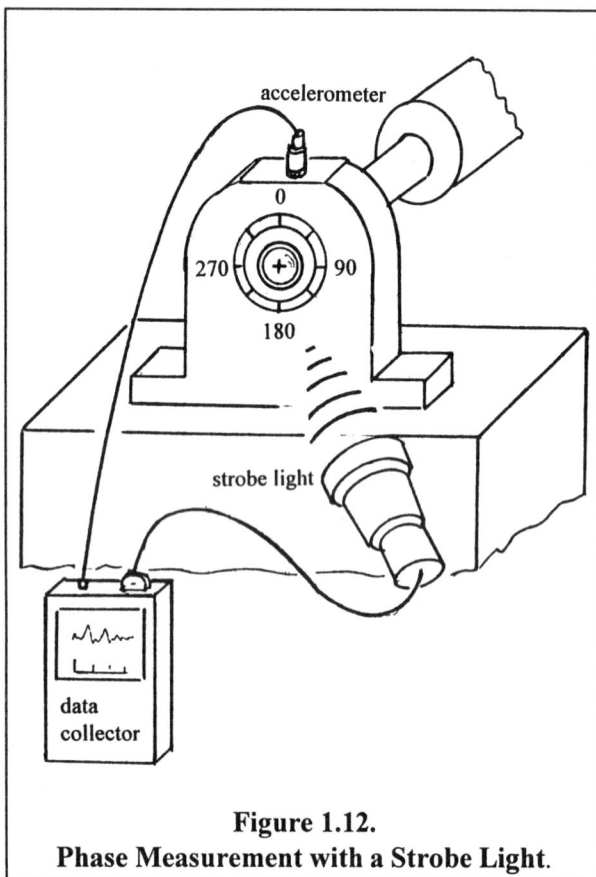

Figure 1.12. Phase Measurement with a Strobe Light.

Phase can also be measured with a strobe light (Figure 1.12). The strobe light is triggered by the vibration signal — that is, when the signal changes from negative to positive voltage, the strobe light flashes — and illuminates a mark on the shaft at some position relative to a protractor. Phase at different positions can be identified by moving the vibration transducer while observing the phase angle.

Vibration Analysis

Periodic motion can be broken down into harmonic motions. The total periodic vibration shown in Figure 1.13 can be represented by the sum of two harmonic vibrations, 1 and 2.

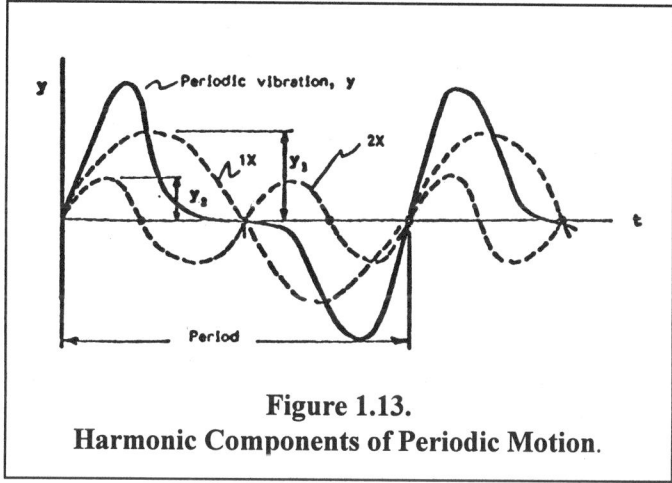

Figure 1.13. Harmonic Components of Periodic Motion.

Note that vibration 2 has a frequency two times that of vibration 1. Vibration 2 is called a second harmonic of vibration 1 because its frequency is exactly two times (2x) that of vibration 1. When the frequency of vibration 1 is equal to the rotational speed (RPM) of the machine vibration, vibration 2 is called a second order.

Periodic motion has a specific shape when the two components are in phase as shown in Figure 1.13. If the phase of the two components is changed, the magnitude of the peak value of vibration — i.e., amplitude — will change. The sum of the two amplitudes is in general not equal to the peak of the periodic waveform. Only when the fundamental frequency (1x) leads the second order component (2x) by 45° or 225° will the sum of the peak amplitudes of vibration 1 and vibration 2 be equal to the peak value of the total vibration. Other phasing will result in a lower total peak amplitude than the sum of the components.

The amplitude and frequency components that make up the time waveform are shown directly in the frequency spectrum of Figure 1.14 (upper plot), which shows amplitude vs frequency. The breakdown of a complex periodic waveform into its components is shown in Figure 1.15. The spectrum shows the breakdown of the waveform into its harmonic components. The amplitudes of the harmonics shown in the spectrum were obtained from a spectrum analyzer. The time domain cannot be reconstructed from the frequency spectrum of this complex periodic waveform unless the phase angles between each harmonic component are known.

An FFT analyzer utilizes a block of data acquired over a period of time and related to a frequency range selected prior to processing. A digital computer containing an algorithm (a defined mathematical procedure) carries out the FFT analysis. The FFT analyzer displays the components of vibration in bins, or lines (typically 400), equally spaced across the frequency range. Bins can be considered a series of filters.

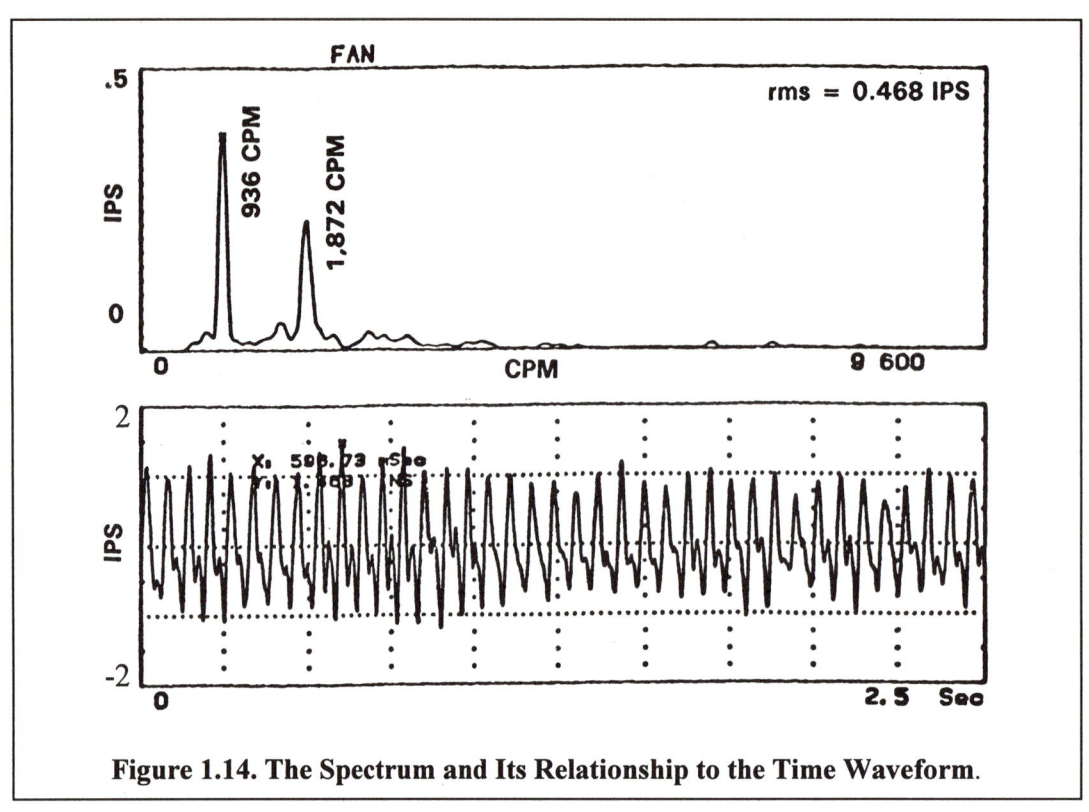

Figure 1.14. The Spectrum and Its Relationship to the Time Waveform.

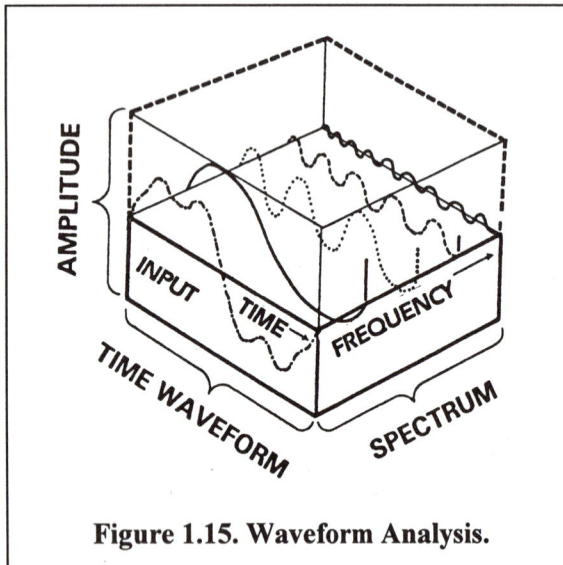

Figure 1.15. Waveform Analysis.

Excitation

The purpose of vibration analysis is to identify defects and evaluate machine condition. Frequencies are used to relate machine faults to the forces that cause vibration. It is therefore important to identify the frequencies of machine components and machine systems before performing vibration analysis. The forces are often the result of defects or wear of components or are due to equipment design or such installation problems as misalignment, soft foot, and looseness. Table 1.3 is a list of some forcing frequencies commonly associated with machines. Because the vibration source is related in some way to operating speed, it is important to identify the operating speed before proceeding with an analysis.

1.14

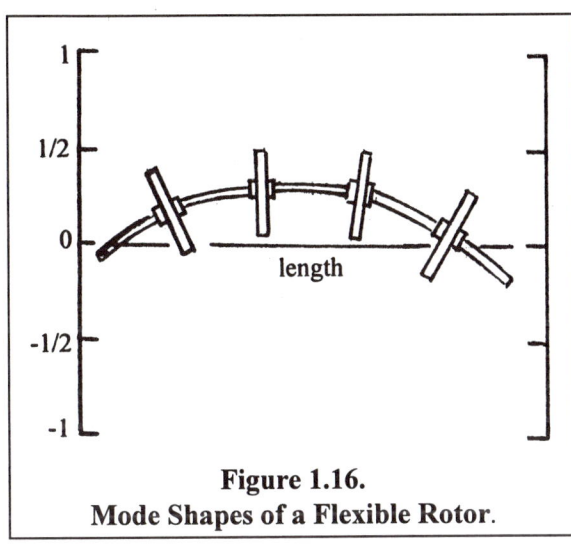

**Figure 1.16.
Mode Shapes of a Flexible Rotor.**

Natural Frequencies, Mode Shapes, and Critical Speeds

Natural frequencies are determined by the design of the machine or a component. They are properties of the system and are dependent on its distribution of mass and stiffness (see Figure 1.2). Every system has a number of natural frequencies. They are not, however, multiples of the first natural frequency (with the exception of rare instances of simple components). Natural frequencies are not important in machine diagnostics unless a forcing frequency occurs at or close to a natural frequency or impacting occurs within the machine. If a forcing frequency is close to a natural frequency, a resonance exists, and the vibration level is high because the machine absorbs energy easily at its natural frequencies. If this forcing frequency is an order of the operating speed of the machine, it is termed a critical speed. Only natural frequencies in the range of forcing frequencies are of interest in the vibration analysis of machines.

Table 1.3. Some Forcing Frequencies Associated with Machines.

Source	Frequency (multiple of RPM)
Fault Induced	
mass unbalance	1x (frequency is once per revolution)
misalignment	1x, 2x
bent shaft	1x
mechanical looseness	odd orders of x
casing and foundation distortion	1x
antifriction bearings	bearing frequencies, not integer ones
impact mechanisms	multi-frequency depending on waveform
Design Induced	
universal joints	2x
asymmetric shaft	2x
gear mesh (n teeth)	nx
couplings (m jaws)	mx
fluid-film bearings (oil whirl)	0.43x to 0.47x
blades and vanes (m)	mx
reciprocating machines	half & full multiples of speed, depending on design

Mode shapes of a system are associated with its natural frequencies. The shape assumed by a system as it vibrates at a natural frequency is called its mode shape. A mode shape does not provide information about absolute motions of the system but consists of deflections at selected points. The deflections are determined relative to a fixed point in the system — usually at the end of a shaft. Absolute motions can be determined only when damping and vibration forces are known. An example of the mode shape of a flexible rotor is shown in Figure 1.16. Rocking modes of rigid rotors are governed by bearing flexibility. Flexible rotors can vibrate in modes with lateral, torsional, and axial motions. The point in a mode shape at which the deflection is zero is called a node. Obviously transducers should never be mounted at a node.

Summary of Basic Vibrations

- Three important characteristics of vibration are frequency, amplitude, and phase.
- The frequency is the number of cycles per unit of time.
- The period is the time required for one cycle of vibration; it is the reciprocal of frequency.
- Amplitude is the maximum value of vibration at a given location on a machine. It is expressed in mils (displacement), in./sec (velocity), or g's (acceleration).
- The amplitude of vibration is expressed in units of peak, peak to peak, or rms.
- Peak and rms are used with velocity and acceleration; mils peak to peak are used with displacement.
- The measures of vibration — displacement (stress), velocity (fatigue), and acceleration (force) — can be converted one to the other if the vibration is a single frequency (harmonic).
- Phase is the time relationship between vibrations and/or forces of the same frequency.
- A force, or excitation, causes vibration. Vibration always lags force in time.
- Vibratory forces arise from process variables, improper design, bad installation, and defects.
- Vibrations are analyzed in the time waveform and the frequency spectrum.
- Natural frequencies are a property of a machine system and depend on mass and stiffness.
- Resonance occurs when a forcing frequency is equal to or close to a natural frequency.
- A critical speed is a special resonance in a rotating machine.
- Vibration is amplified at resonance.

CHAPTER II
DATA ACQUISITION

Decisions are only as good as the facts they are based on.

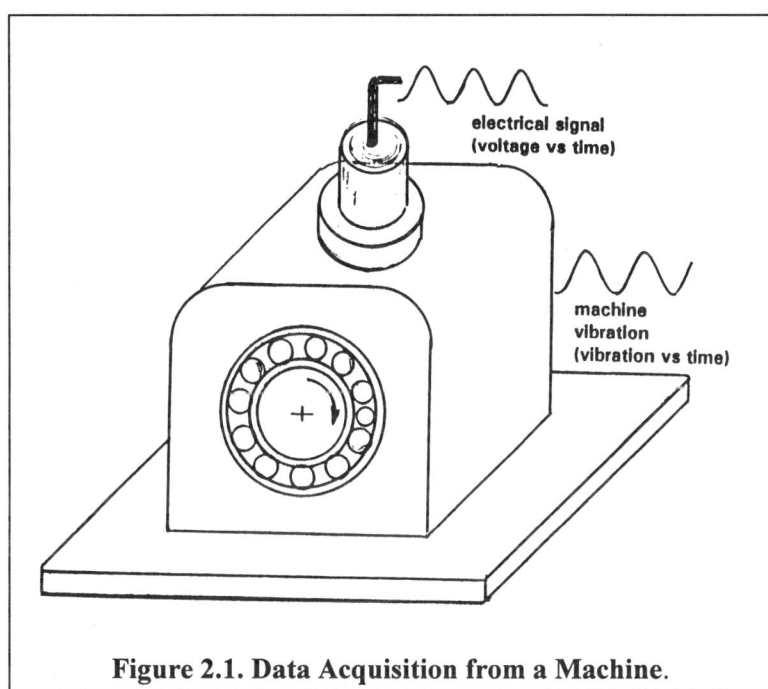

Figure 2.1. Data Acquisition from a Machine.

Vibration data are acquired from a machine by a transducer that converts the mechanical vibration to an electrical signal in volts (Figure 2.1). The quality of the signal obtained from the machine depends on the transducer selected as well as the way in which it is mounted and where it is located.

The acquisition of proper vibration data is the key to effective machine monitoring, fault diagnosis, condition evaluation, and acceptance testing. Quality data acquisition requires planning involving the machine, the nature of the expected vibration data, available instrumentation, and the purpose of the test. Prior to data acquisition the analyst must formulate a technically sound and cost-effective plan based on the purpose of the data (i.e., monitoring, diagnostics, condition evaluation, or acceptance testing). The topics considered in this chapter include selection of the measure (displacement, velocity, or acceleration), the transducer, and the transducer mounting and location. Because all data are digitized prior to storage, acquisition times and sample sizes must also be considered when data are acquired so that proper displays can be provided for analysis and evaluation. The key to quality work in the vibration field is proper data acquisition.

Selecting a Measure

A measure is a unit or standard of measurement that provides a means for evaluating data. Three measures of vibration are available — *displacement*, *velocity*, and *acceleration*. Ideally the transducer would directly provide the selected measure. Unfortunately, transducer limitations do not always allow direct measurement of vibration in the proper measure.

The measure is selected on the basis of the frequency content of the vibration present, the design of the machine, the type of analysis to be conducted (e.g., faults, condition, design information), and the information sought. *Absolute displacement,* which is used for low-frequency structural vibration (0 to 20 Hz), relates to stress (shaft or structure) and is typically measured with a double integrated accelerometer. Absolute displacement of a shaft must be measured with either a contacting transducer or a noncontacting transducer in combination with a seismic transducer. Unfortunately, frequency must also be considered when the severity of displacement and acceleration are assessed. *Relative shaft displacement*, which is measured with a proximity probe, shows the extent of bearing clearance taken up by vibration and is used over a wide frequency range.

For general machinery monitoring and analysis in the span from 10 Hz to 1,000 Hz, *velocity* is the default measure. Velocity as a time rate of change of displacement is dependent upon both frequency and displacement and relates to fatigue. It has been shown to be a good measure in the span from 10 Hz to 1,000 Hz because a single value for rms or peak velocity can be used in rough assessments of condition without the need to consider frequency. Most modern data collectors use accelerometers; the signal must be integrated to obtain velocity.

Acceleration is the measure used above 1,000 Hz; it relates to force and is used for such high-frequency vibrations as gear mesh and rolling element bearing defects. Acceleration and velocity are absolute measures taken on the bearing housing or as close to it as possible. Relative displacement between the machine housing and the rotor is typically measured by a permanently mounted proximity probe.

Some general applications of measures and their applicable frequency spans are given in Table 2.1. Default frequency spans for data collectors are shown in Table 2.2. Various machine-specific measures are listed in Table 2.3. Selection of a measure — displacement, velocity, acceleration — for evaluating faults and conditions of machines is therefore based on the useful frequency span of the vibration measure (Table 2.1), the default frequency spans (Table 2.2), and the application (Table 2.3).

Example 2.1: Select a measure or measures for the 9 Mw single-reduction gearbox described in Table 2.3. The gearbox uses fluid-film bearings and is sufficiently large (greater than 500 HP) to justify permanently-mounted proximity probes for evaluating the position of the journal in the bearing and the ratio of the vibration to the bearing clearance. The analyst is thus able to access severity of the shaft vibration of the journal. Because the gear-mesh frequency (3,000 Hz) is greater than 1,000 Hz (see Table 2.1), casing acceleration must be monitored and analyzed. Frequency spans of 10,000 Hz (see Table 2.2) should be monitored in acceleration and 75,000 CPM (high-speed input) and 12,000 CPM (output) respectively in shaft vibration from the proximity probes.

Example 2.2: Select measure(s) for a low-speed 300 RPM dryer roll (see Table 2.3). The multi-ton roll is mounted on large rolling element (26) bearings. Because the roll operates at such a low speed, mass unbalance is not a major consideration since the force is small. The highest rolling element bearing frequency is the ball pass frequency of the inner race. It can be estimated as

$$BPFI = (0.6)(RPM)(N)$$
$$BPFI = (0.6)(300)(26) = 4,680 \text{ CPM } (78 \text{ Hz})$$

Therefore, the frequency span is 780 Hz (see Table 2.2). This value is within the velocity range (see Table 2.1).

Table 2.1. Machine Vibration Measures.

Measure	Useful Frequency Span	Physical Parameter	Application
relative displacement	0-1,000 Hz	stress/motion	relative motions in bearings/casings
absolute displacement	0-10 Hz	stress/motion	machine and structural motion
velocity	10-1,000 Hz	energy/fatigue	general machine condition, medium-frequency vibrations
acceleration	>1,000 Hz	force	general machine condition, medium-/high-frequency vibrations

Example 2.3: Select measure(s) for a 200 HP-four pole induction motor with eight rolling elements in the bearings. The operating speed vibrations have a frequency of 1,800 CPM (30 Hz) and a frequency span of 300 Hz, which is within the velocity range. The bearing frequency span is

$$(BPFI)(10) = (0.6)(8)(1,800)(10) = 86,400 \text{ (1,440 Hz)}$$

Because the majority of the activity is in the velocity range, a velocity transducer can be used even though some activity is above 1,000 Hz. The useful frequency spans of all measures overlap. Therefore, the measure should be selected from the predominant portion of the frequency activity of the component. For example, if the default frequency span for the bearing had been 2,880 Hz (16 rolling elements), acceleration would have been selected as the measure for the bearings. Unfortunately, the shaft vibration frequency span of 300 Hz remains within the velocity range. Therefore, two measures, velocity and acceleration, are required.

Table 2.2. Default Frequency Spans for Data Collectors.

Component	Span
shaft vibration	10 x RPM
gearbox	3 x GM
rolling element bearings	10 x BPFI
pumps	3 x VP
motors/generators	3 x 2 LF
fans	3 x BP
sleeve bearings	10 x RPM

Vibration Transducers

Information about vibration is acquired by transducers positioned at optimal locations within a machine system. Transducers convert mechanical vibrations to electronic signals that are conditioned and processed by a wide variety of instruments. These instruments provide the information necessary to monitor machine condition, verify performance, diagnose faults, and identify parameters. Magnitude, frequency, and phase between two signals are used for evaluation. Transducer selection is based on sensitivity, size required, selected measure, frequency response, and machine design and speed.

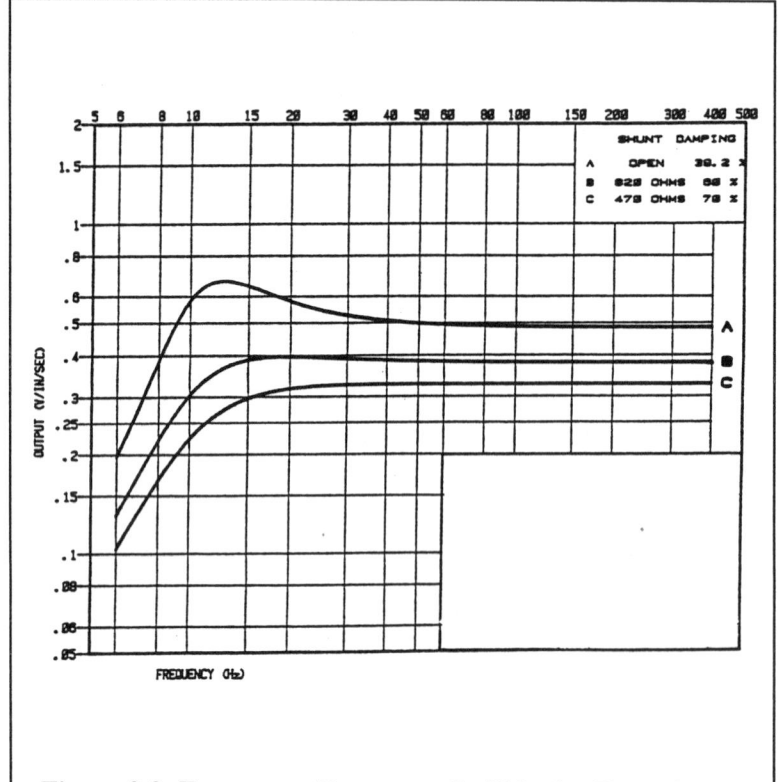

Figure 2.2. Frequency Response of a Velocity Transducer.

The response of any instrument, including transducers, determines how well the instrument responds to a stimulus (voltage or vibration) at a given frequency. Analysts want a flat frequency response at all frequencies. Does the transducer provide an electrical signal that is proportional to the vibration it is measuring? Unfortunately, no is sometimes the answer. In Figure 2.2, for example, the frequency response curve for a velocity transducer is not flat. At the lower frequencies it rolls off — that is, it responds less to the same strength signal than it does at frequencies greater than 20 Hz. This

Table 2.3. Measures for Selected Machines.[1]

Machine	HP/Mw	Speed (RPM) Frequencies (Hz)	Bearing Type	Measure(s)	Transducer
Gearbox - single reduction	9 Mw	7,500 RPM input 1,200 RPM output GM = 3,000 Hz	fluid-film	displacement-shaft[2] acceleration-casing[3]	proximity probe accelerometer
Gearbox - double reduction	400 HP	1,800 RPM input 200 RPM output GM = 375,725 Hz	15 rolling elements	acceleration velocity	accelerometer integrated accelerometer or velocity
Steam Turbine	18,000 HP	5,000 RPM	fluid film	displacement-shaft	proximity probe
Steam Turbine	500 Mw	3,600 RPM	fluid-film	displacement-shaft	proximity probe
Gas Turbine	50 Mw	9,000 RPM	fluid-film rolling element	displacement-shaft acceleration-casing	proximity probe accelerometer
Large Induction Motor	4,000 HP	3,600 RPM	fluid-film	displacement-shaft	proximity probe
Induction Motor	200 HP	1,800 RPM	8 rolling elements	velocity-casing	integrated accelerometer or velocity
Diesel Engine	400 HP	1,800 RPM	fluid film	velocity-casing	integrated accelerometer or velocity
High-Performance Centrifugal Pump	18,000 HP	5,000 RPM	fluid-film	displacement-shaft velocity-casing	proximity probe integrated accelerometer or velocity
Centrifugal Pump	200 HP	1,800 RPM	12 rolling elements	velocity-casing	integrated accelerometer
Reciprocating Pump	200 HP	300 RPM	15 rolling elements	velocity-casing	integrated accelerometer or velocity
Centrifugal Compressor	1,000 HP	5,000 RPM	fluid-film	displacement-shaft	proximity probe
Reciprocating Compressor	500 HP	480 RPM	fluid-film	velocity-casing	integrated accelerometer or velocity
Dryer Roll		300 RPM	26 rolling elements	velocity-casing	integrated accelerometer or velocity

[1] conventional measures only; HFD= envelope detection, and other special techniques not included
[2] shaft = relative shaft vibration
[3] casing = bearing cap

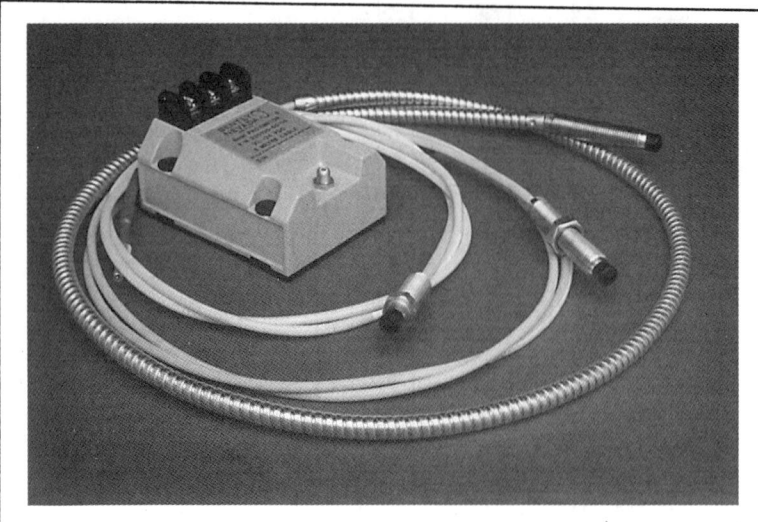

Figure 2.3. Proximity Probe.
Courtesy of Bently Nevada Corporation

means that, if the same scale factor of 484 mv/IPS is used throughout the entire frequency range, the amplitudes at the lower frequencies will be lower than their actual values. Frequency response (amplitude versus frequency) curves such as the one shown in Figure 2.2 are typically supplied by the transducer manufacturer.

The sensitivity of the transducer is dependent upon its voltage output for a given vibration input; for example, 200 mv/mil, 500 mv/IPS, 100 mv/g. The higher the voltage output per engineering unit, the more sensitive the transducer.

Proximity probes. The proximity probe (noncontacting eddy current displacement transducer) shown in Figure 2.3 measures static and dynamic displacement of a shaft relative to the bearing housing. It is permanently mounted on many machines for monitoring (protection) and analysis. Applications of the probe to vibration measurements in radial and axial positions are covered in detail in API 670 [1].

The probe is a coil of wire surrounded by a nonconductive plastic or ceramic material contained in a threaded body. An oscillator-demodulator, often referred to as a driver or proximitor, is required to excite the probe at about 1.5 mega Hertz (MHz). The resulting magnetic field radiates from the tip of the probe. When a shaft is brought close to the probe, eddy currents are induced that extract energy from the field and decrease its amplitude. This decrease in amplitude provides an AC signal directly proportional to vibration (mv/mil). The DC voltage from the oscillator-demodulator varies in proportion to the distance between the probe tip and the conducting material. The sensitivity of the probe is generally 200 mv/mil (8 mv/μm) with a gap range (distance from probe tip to shaft) from 0 to 80 mils. The oscillator-demodulator requires a supply of negative 24 v DC. The probe must be shielded and grounded.

Figure 2.4. Velocity Transducer.

Velocity transducers. The velocity transducer (Figure 2.4) is a seismic transducer (i.e., it measures absolute vibration) that is used to measure vibration levels on casings or bearing housings in the range from 10 Hz to 2,000 Hz. The transducer is self excited — that is, it requires no power supply — and consists of a permanent magnet mounted on springs encased in a cylindrical coil of wires. Motion of the coil relative to the magnet generates a voltage proportional to vibration velocity. The self-generated signal can be directly passed to an oscilloscope, meter, or analyzer for evaluation. A typical velocity transducer generates 500 mv/(in./sec) except at frequencies below 10 Hz (see Figure 2.2), which is the natural frequency of the active element. The reduction of output below 10 Hz requires that a frequency-dependent compensation factor be applied to the amplitude of the signal. The measured phase also changes with frequency at frequencies below 10 Hz. The velocity transducer can be used to evaluate vibration velocity in order to assess machine condition when the frequency range of concern is within the flat frequency response (10-2,000 Hz) of the transducer. Velocity transducers can be used to measure shaft vibration with a fish tail, a simple wooden device that attaches to the transducer. A vee notch permits the fish tail to ride on the rotating shaft. Keys and other variations of the shaft surface pose safety hazards.

Accelerometers and other force transducers. Accelerometers are used to measure vibration levels on casings and bearing housings; they are the transducers typically supplied with electronic data collectors. An accelerometer (Figure 2.5) consists of a small mass mounted on a piezoelectric crystal that produces an electrical output proportional to acceleration when force is applied in the form of a vibrating mass. Force transducers such as modal hammers and force gauges (Figure 2.6) also contain a piezoelectric crystal, but the electrical output of the crystal is proportional to the force applied.

The piezoelectric crystal generates a high impedance signal that must be modified by charge or voltage conversion to low impedance. The size of an accelerometer is proportional to its sensitivity. Small

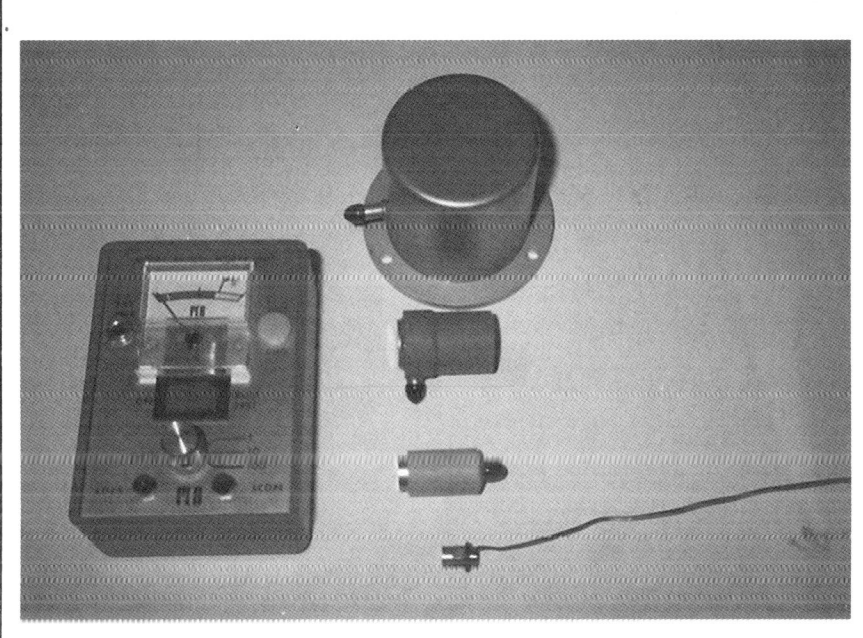

Figure 2.5. Accelerometers and Power Supply.
Courtesy of PCB Piezotronics Inc.

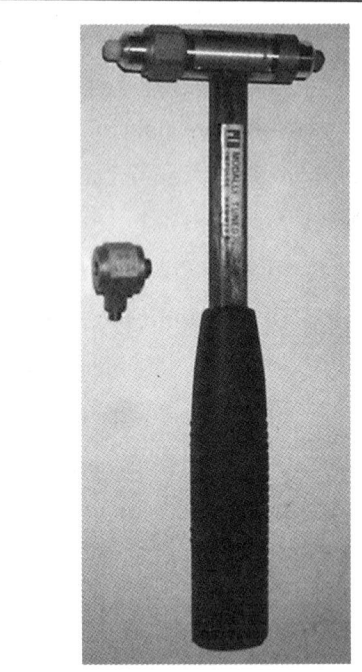

Figure 2.6. Modal Hammer and Force Gauge.
Courtesy of PCB Piezotronics Inc.

accelerometers (the size of a pencil eraser) have a sensitivity of 5 mv/g (1 g = 386.1 in./sec^2) and a flat frequency response to 25 kHz. A 1,000 mv/g accelerometer, which is used for low-frequency measurement, may be as large as a velocity transducer; however, the limit of its usable frequency span may be to 1,000 Hz. The analyst should be aware of the properties of each accelerometer being used.

If vibration velocity is desired, the signal is usually integrated before it is recorded or analyzed; an analog integrator/power supply is shown in Figure 2.7. This device has its own frequency response characteristics and rolls off at low frequencies. Accelerometers are recommended for permanent seismic monitoring because of their extended life [2] and because their cross sensitivity is low. (Cross sensitivity means that the transducer generates a signal in direction X from vibration in direction Y.) However, cable noise, transmission distance, and temperature sensitivity of the accelerometer must be carefully evaluated. Excellent guidelines are available from vendors for accelerometer use.

Triggering Devices

When it is advantageous to directly associate vibration data to a rotating shaft or other vibrating object, a triggering device is used. This device senses or is timed to the frequency of a mark, indentation, or protrusion of the rotating shaft and sends a signal to an analyzer or oscilloscope that initiates data acquisition. Thus, data are acquired at the same point on the shaft each time the trigger sends a signal to the analyzer. The frequency of the signal from the triggering device is associated with shaft speed or some multiple of it; the phase between the reference and the vibration signal can be obtained (see Figure 1.11).

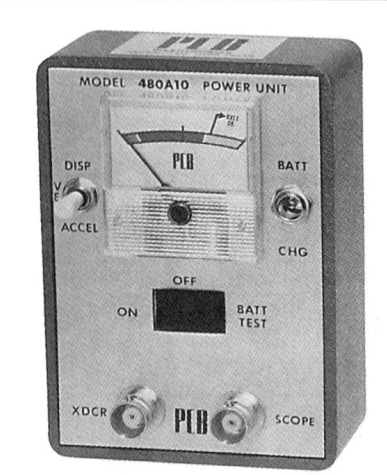

Figure 2.7. Analog Integrator and Power Supply.
Courtesy of PCB Piezotronics Inc.

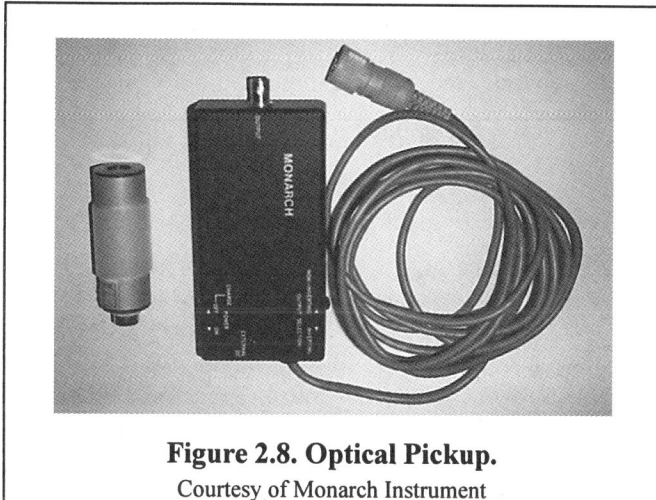

Figure 2.8. Optical Pickup.
Courtesy of Monarch Instrument

Optical pickups. The optical pickup (Figure 2.8) is most often used to obtain the once-per-revolution reference signal required to measure the phase angle between a piece of reflecting tape on a shaft and a once-per-revolution vibration peak (see Figure 1.11). When energized by light pulses from the reflecting tape, the pickup sends a voltage pulse to the analyzer. The analyzer can compare the timing of the tape (shaft reference pulse) to other events — i.e., other marks on the shaft, vibration peaks — or its own readings (to determine shaft speed).

Optical pickups can also be used to observe time elapsed between equally-spaced marks on a rotating shaft when torsional vibration measurements are made. The optical system includes a pickup mounted adjacent to the shaft, reflective tape on the shaft, and a power supply/amplifier.

Magnetic pickups and proximity probes. The magnetic pickup (Figure 2.9), which is self-excited, can be used as a triggering device because a voltage pulse occurs when the pickup encounters a discontinuity such as a keyway. The pickup is placed about 20 mils from the shaft. Magnetic pickups are used in torsional vibration measurements to produce series of pulses proportional to shaft speed. If torsional vibrations are present, the time between pulses varies, producing frequency modulation.

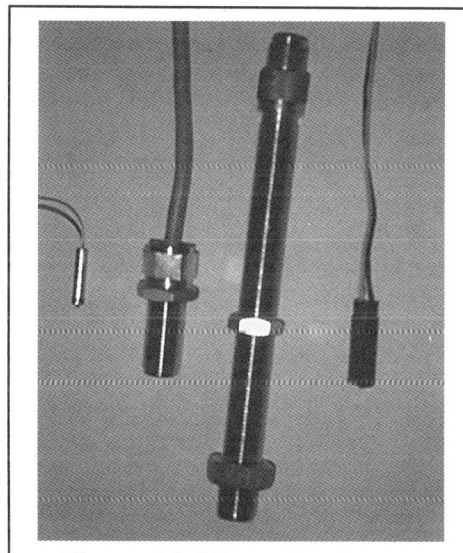

Figure 2.9. Magnetic Pickups.

One disadvantage of the magnetic pickup is that signal conditioning is sometimes difficult because the magnitude of the voltage is dependent on speed. The proximity probe, which is powered, provides the same triggering function without this disadvantage.

Strobe light. The strobe light is used to measure speed or phase in conjunction with a vibration sensor. To measure speed, the frequency at which the light flashes is tuned to shaft speed by selecting a mark on the shaft and adjusting the flash frequency of the strobe until the mark is stationary. To measure phase, the strobe light is fired by the vibration signal when it crosses from minus to plus. This means that the high spot will always lead the measured phase angle by 90°.

Transducer Selection

Important considerations in transducer selection include frequency response, signal-to-noise ratio, the sensitivity of the transducer, and the strength of the signal being measured. The frequency range of the transducer must be compatible with the frequencies generated by the mechanical components of the machine. Otherwise, another transducer must be selected and the signal converted to the proper measures. For example, if the velocity measure is desired at frequencies above 2,000 Hz, an accelerometer integrated to velocity should be selected to obtain the signal. If the time waveform of the velocity measure is desired, the signal must be acquired from a velocity pickup or analog integrated signal from an accelerometer, either within or external to the data collector.

A single transducer — usually an accelerometer because of its small size and frequency characteristics — is provided with most electronic data collectors. The frequency response characteristics of the unit must be assessed so that the user will not try to detect vibrations to which the collector does not respond. For example, if a typical collector with an accelerometer is set up to respond to frequencies up to 8 kHz and a gearbox has gear mesh at 10 kHz, the signal will be dropped out. Acceleration is measured, and most collectors provide readouts in acceleration or velocity. The parameter selected depends on the criteria chosen.

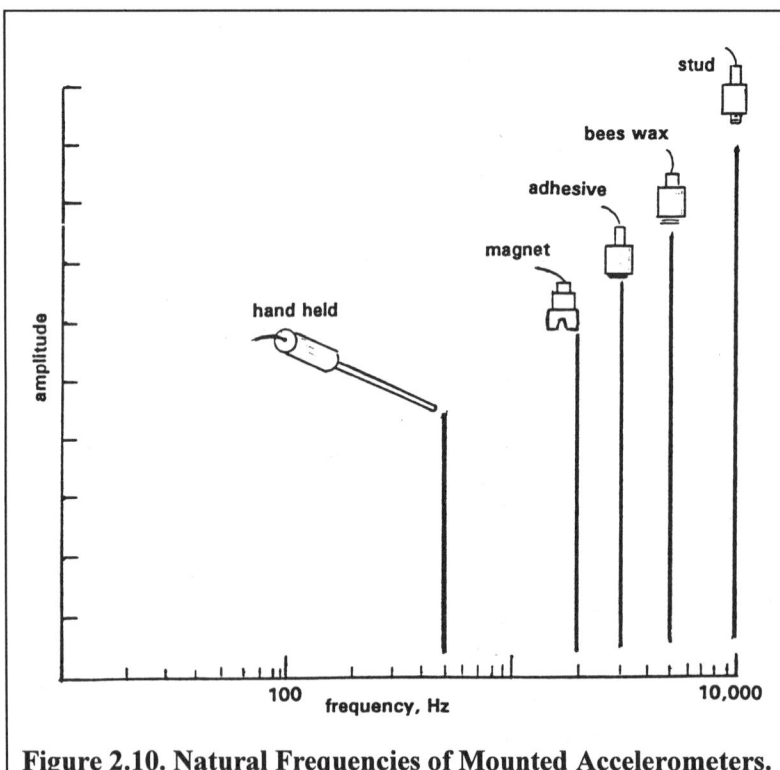

Figure 2.10. Natural Frequencies of Mounted Accelerometers.

The cable that transmits the signal to the data collector can cause erroneous readings. Many standard cables are specially wound cords that are more convenient than the standard coaxial construction. But, because many conductors are flexible at the core, individual strands may fail at stress points as a result of handling or packing in a carrying case. In addition, the terminals must be handled carefully. Many accelerometer manufacturers issue Amphenol 97 Series™ connectors. They have threaded assemblies that can loosen in the field, causing the conductors to twist and break. A fail-safe approach it to apply Loctite™ to all threaded connections when the plug is new. A spare cable can be helpful in resolving questions about the integrity of one currently in use. These cables are computer connectors and must be handled with care.

Transducer Mounting

The method used to mount a vibration transducer can affect the frequency response curve because the natural frequency of an accelerometer can decrease, depending on the mounting method used — hand-held, magnetic, adhesive, threaded stud (Figure 2.10). The mounting method chosen should provide flat frequency response throughout the frequency range being studied. Data on transducer mounting are available [3]. It can be seen from Table 2.4 that a stud mount on a clean flat surface with a good finish provides the highest frequency response. The response decreases progressively for adhesive and magnetic mountings. Reliability is lowest with a hand-held accelerometer. The values in Table 2.4 are intended to serve as guidelines. Each accelerometer and its mounting have a unique natural frequency and therefore usable frequency range.

Table 2.4. Approximate Frequency Spans for 100 mv/g Accelerometers.

Method	Frequency Limit
Hand Held	500 Hz
Magnet	2,000 Hz
Adhesive	2,500-4,000 Hz
Bees Wax	5,000 Hz
Stud	6,000-10,000 Hz

Transducer Location

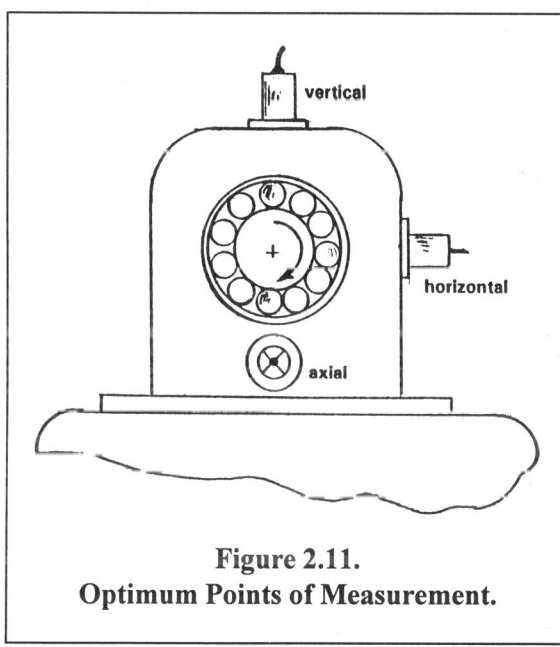

Figure 2.11. Optimum Points of Measurement.

The key to accurate vibration measurement is placement of the transducer at a point that is responsive to machine condition. In any event the transducer should be placed as close to the bearing as is physically possible. Figure 2.11 shows the optimum points for mounting transducers for data acquisition. The horizontal and vertical locations at the bearing centerline are shown. These locations are used to sense the vibrations from radial forces such as mass unbalance. Vibrations from axially-directed forces are measured in the axial direction in the load zone. In the figure the weight of the rotor causes the load zone to be at six o'clock.

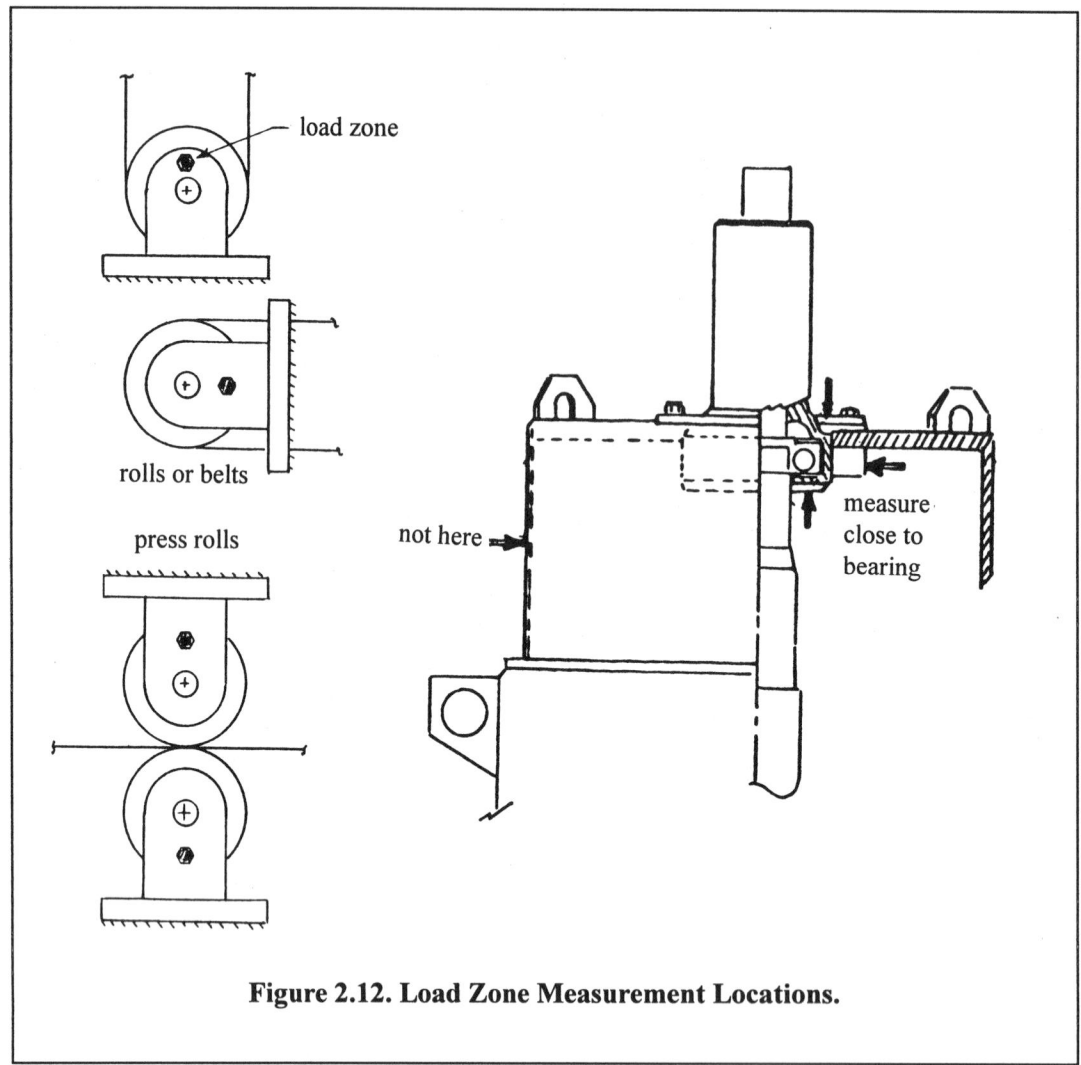

Figure 2.12. Load Zone Measurement Locations.

The transducer must be placed as close to the bearing as possible, even though placement is restricted by such components as housings, coupling guards, and fan covers. Recommended locations for placing transducers on a typical machine are shown in Figure 2.12. Internal design details are necessary on complex housings to determine bearing type and the optimal path for transmission of the mechanical signal. When bearings are inaccessible, transducers can be mounted and hard wired to a bulkhead fitting to allow uninterrupted access to the measurement point.

The existing internal diametral clearance of slow-speed, heavily-loaded bearings — typically spherical roller bearings — allows for inner ring/rolling element/outer ring to contact only at the load zone of the assembly.

In general, radial readings are taken on radial bearings; that is, any antifriction bearing with a contact angle of 0°. Radial bearings are used in electric motors, in medium- to light-duty fans, and in power transmission units not subject to axial loading.

Angular contact bearings or any bearing absorbing thrust have a radial-to-axial coupling that requires an axial measurement for accurate condition monitoring. Gears other than spur gears absorb thrust and should be measured in the axial direction. Radial measurements are required to ascertain conditions at operating speed (1x) such as unbalance, misalignment, looseness, resonance, and bent shaft.

Frequency Spans

Spectra can be collected as part of the screening function on most data collectors. The frequency spans must properly reflect the sample. And the proper transducer must be selected. Geared units may generate mesh frequencies with significant harmonics that are clipped by the 2 kHz roll off of a velocity transducer. Therefore, measurement must be based in acceleration. Clipping can also occur when the range of the spectrum is less than the maximum frequency being transmitted. Table 2.2 contains recommended frequency spans for spectra taken from rotating machines for monitoring and analysis. The spans are based on RPM and other machine frequencies. Clipping in the frequency span is indicated by spectral energy values that are significantly lower than overall levels.

However, if the spectral span is broad, resolution can be reduced to the point that no discrete frequency information is available. If adequate resolution is not available from the default frequency spans, multiple data samples must be acquired and analyzed. An optimum configuration allows sufficient resolution to analyze the operating speed frequency and sidebands as well as the range to measure higher bearing or gear-mesh frequencies. It is desirable to split the data point into two or three spans or to increase lines of resolution to obtain better resolution. Therefore, either several data acquisition cycles may be necessary at the same test point or the data collector must be capable of processing the data in several spans from a single sample.

Data Display

Vibration data from a machine running at a constant operating speed are generally repetitive. Small variations do occur as a result of the influence of load, temperature, and process. Environmental and load conditions should be noted when the data are taken. Data are typically displayed in the spectrum, time waveform, and orbit.

> **Example 2.4**: A grinding spindle that operates at 6,000 RPM is supported on rolling element bearings (19 elements). The recomended frequency spans (Table 2.2) for this machine are 60,000 CPM (1,000 Hz) for shaft vibration and (19) (0.6) (6,000) (10) = 684,000 CPM (11,400 Hz) for the rolling element bearings; most likely a frequency span of 12,000 Hz will be available on the data collector.
>
> This situation requires a velocity spectrum with a span of 1,000 Hz for operating speed faults. If a 400-line spectrum is used with a Hanning window, a resolution of (1,000 Hz/400 lines) (3) = 7.5 Hz (see *Chapter III*) is obtained. This resolution is adequate for operating speed faults.
>
> For the rolling element bearings, the lowest sideband is 0.4 RPM *(see Chapter IV)* = 2,400 CPM (40 Hz). Therefore, a minimum frequency resolution of 40 Hz is required. Thus, the number of lines required with a Hanning window is
>
> $$N = (F_{max}/RES) (3) \text{ (Chapter III)} = (12,000/40) (3) = 900 \text{ lines}$$
>
> Therefore, 1,600 lines of resolution are required.

Time waveform. The time waveform is a plot of vibration amplitude versus time. It reflects the physical behavior of the machine in the vibration signal — particularly if it is displayed on an analog oscilloscope, because no compromises are made. A time waveform and spectrum from a fan are shown in Figure 1.14. The time waveform is useful in identifying unique events in a machine and the rate at which they are repeated. The length (in seconds) of the display of data from the time waveform depends on the information sought. It is typically related to the operating period τ of the machine; τ in seconds is equal to 60/RPM. The best phase angle resolution for basic balancing is obtained by using the fundamental period τ for display. In Figure 2.13 the shape of the time waveform is obtained from a 12τ display (400 msec display/33.3 msec fundamental period). Figure 1.14 has a time waveform with a 39τ display (60/936 = 0.064 sec = τ; span = 2.5 sec. Therefore, 2.5/0.0641 = 39, which allows the analysis of events over a long period of time.

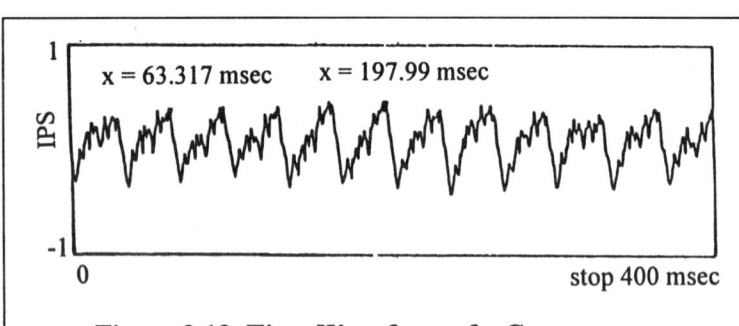

Figure 2.13. Time Waveform of a Compressor.

Spectrum. The setup of the spectrum (Figure 1.14) is determined by the frequency span of the data in order that all information will be obtained. Resolution, dynamic range, and amplitude accuracy are determined by the setup of the FFT analyzer (*Chapter III*). In the Figure the response at operating speed (1x) of the shaft to vibration is being analyzed. A 10x frequency span was thus selected. These data were processed in a fixed 400-line analyzer. Thus, no variability is allowed on resolution except for the window. A flat top window (*Chapter III*) was used so that the correct amplitude would be obtained, but resolution was sacrificed (183.2 CPM for flat top vs 72 CPM for Hanning window; *see Chapter III*). In such a situation, if the frequency span and better resolution are needed, two or more spectra should be processed in different frequency spans. With a data collector the analyst has the option of increasing the lines of resolution instead of taking more spectra.

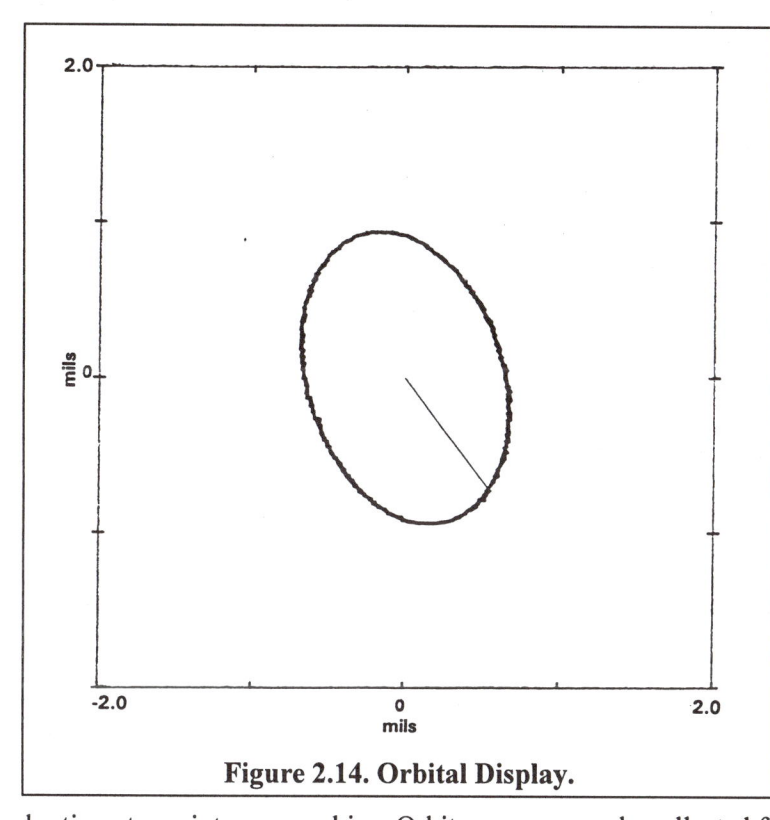

Figure 2.14. Orbital Display.

Orbit. The orbit shown in Figure 2.14 is a two-dimensional display of the vibration at a point on a machine. Orbits are commonly collected from proximity probes, which show the physical motions of the shaft with respect to the bearing. Orbits are useful in showing the motions of pedestals, piping, or any structure when a better visualization of the vibrating object is desired.

Summary of Data Acquisition
- Measures — displacement, velocity, or acceleration — to evaluate machine condition are machine specific.
- Vibration transducers should be selected for frequency response, signal strength, size, machine type, and bearing type.
- Signal strength depends on the measure selected and the frequency of interest.
- Acceleration signals are small in magnitude at low frequencies, as are displacement signals at high frequencies.
- Integration of acceleration signals can cause large-magnitude low-frequency noise.
- Frequency response is the ability of a transducer to reproduce the magnitude of vibration within a given frequency range.

- Vibration transducers should be located close to the bearing and mounted to acquire data at the frequency of interest.
- Vibration at operating speed (1x), such as mass unbalance, misalignment, and looseness, are monitored in the radial direction and analyzed in the axial and radial directions.
- Rolling element bearing and gear-excited vibrations are measured in the axial direction.
- Select proper frequency spans and lines so that all vibration activity is captured with adequate resolution.
- Set up the data collector for acquisition to provide data displays that enhance analysis.

References

2.1. API 670, 1986, *Vibration, Axial Position, and Bearing Temperature Monitoring System*, 2nd ed., American Petroleum Institute, Washington, D.C.

2.2. API 678, 1981, *Accelerometer-Based Vibration Monitoring System*, API, Washington, D.C.

2.3. Crawford, A.R. and Crawford, S., *The Simplified Handbook of Vibration Analysis*, Volume One, Computational Systems, Inc. (1992).

CHAPTER III
DATA PROCESSING

When the problem is difficult, the difference between success and failure will be the quality of the processed data.

This chapter deals with the setup and limitations of the instrumentation used to make routine vibration measurements. Included are oscilloscopes, FFT analyzers (fast Fourier transform analyzers), and electronic data collectors. Analog and digital oscilloscopes show the time waveform and are used to screen and analyze its shape and frequencies. Oscilloscopes can be used to evaluate phase and orbits. The FFT analyzer and the electronic data collector are used in spectral analysis and to evaluate the time waveform.

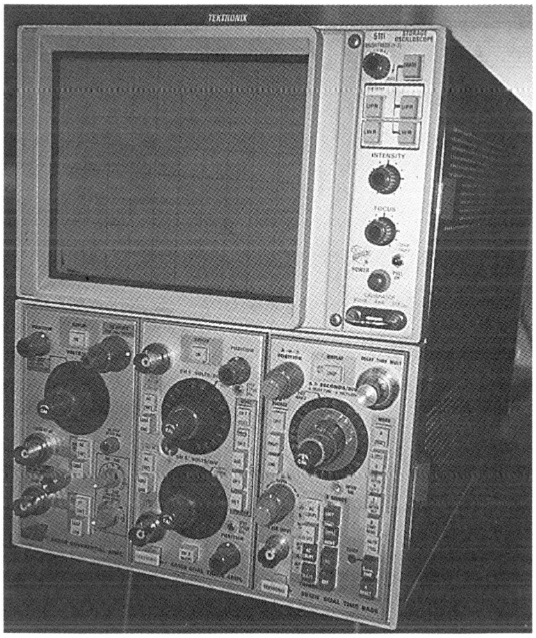

Oscilloscopes

The oscilloscope (Figure 3.1) measures and displays voltages that vary with time. A transducer converts the mechanical vibration into a proportional electrical signal (see Figure 2.1) calibrated in engineering units (EU) of millivolts (mv) per mil, inches per second, or g. Oscilloscopes are used to display time waveforms, orbits, and markers that relate to events such as shaft rotation (Figure 3.2).

Figure 3.1. Analog Oscilloscope.

Triggering. Triggering is an important function of both the oscilloscope and the FFT analyzer. A trigger initiates data acquisition at a specified time or amplitude and controls data acquisition by a specified signal (vibration or trigger). The oscilloscope can be placed on auto trigger to continuously sample data. Triggering can be done on a selected signal based on slope and/or magnitude of voltage. The oscilloscope can be set to make a single sweep that is triggered

Figure 3.2. Oscilloscope Analysis.

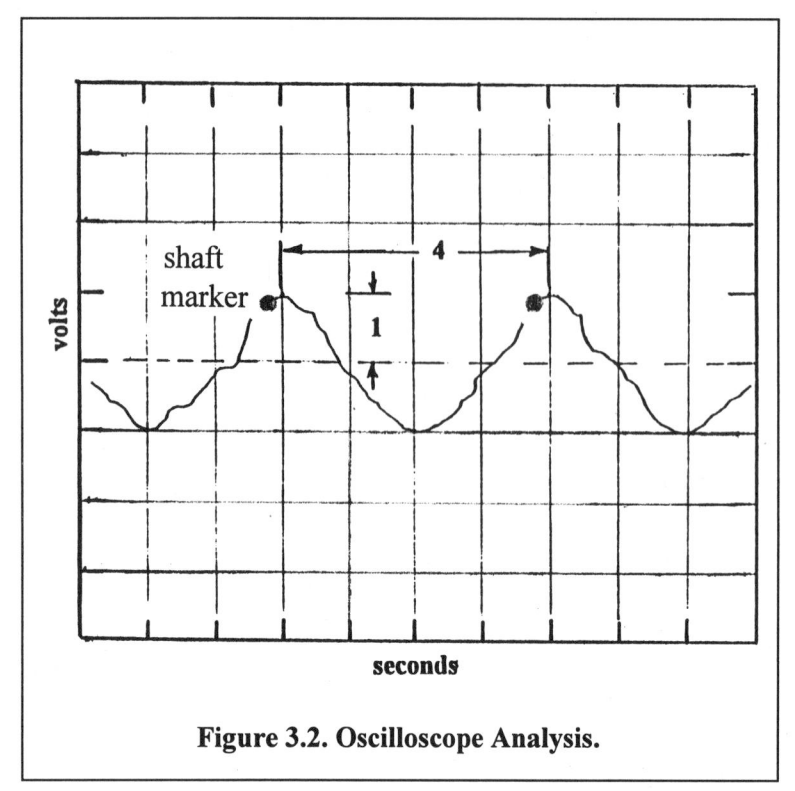

on a specified voltage. For continuous measurement, an optical pickup or a proximity probe can be used as a continuous trigger at shaft frequency.

External intensity input. A controlled blanking or intensity mark can be displayed on the trace of the oscilloscope by applying a +/- five volt signal to the z axis connector. The input can be AC coupled (i.e., no DC passes); if it is not, a capacitor must be use with proximity probes that have more than 5 volts DC. The z axis blanking is used to measure phase and to relate a mark on the shaft to vibration during balancing. The blanking signal is obtained from a triggering proximity probe or optical pickup.

Vertical amplifier. Vertical amplifiers accept the time-varying voltage from a transducer. The controls are calibrated in millivolts (mv)/division (div). The amplitude of the signal in number of divisions is obtained from the screen. Voltage is calculated by multiplying the number of divisions by the amplifier setting in mv/div to obtain millivolts. Vibration amplitude is calculated by dividing millivolts by transducer calibration mv/EU. EU can be mils, IPS, g's, or degrees. The screen has eight divisions vertically (Figure 3.2). Example 3.1 illustrates the use of an oscilloscope to measure amplitude and frequency.

Time base (horizontal amplifier). The primary function of the horizontal amplifier is as a time base. The number of divisions for the period of the signal is obtained from

---

Example 3.1: Find the amplitude and period of the time waveform shown in Figure 3.2.

time base setting = 10 msec/div
amplifier setting = 0.2. volt/div
transducer calibration = 1,000 mv/in./sec

period = (4 div) (10 msec/div) = 40 msec = 0.04 sec

frequency = 1/0.04 sec = 25 Hz = 1,500 CPM

amplitude = (1 div) 0.2 volt/div = 0.2 volt =
200 mv/1,000 mv/in./sec = 0.2 in./sec

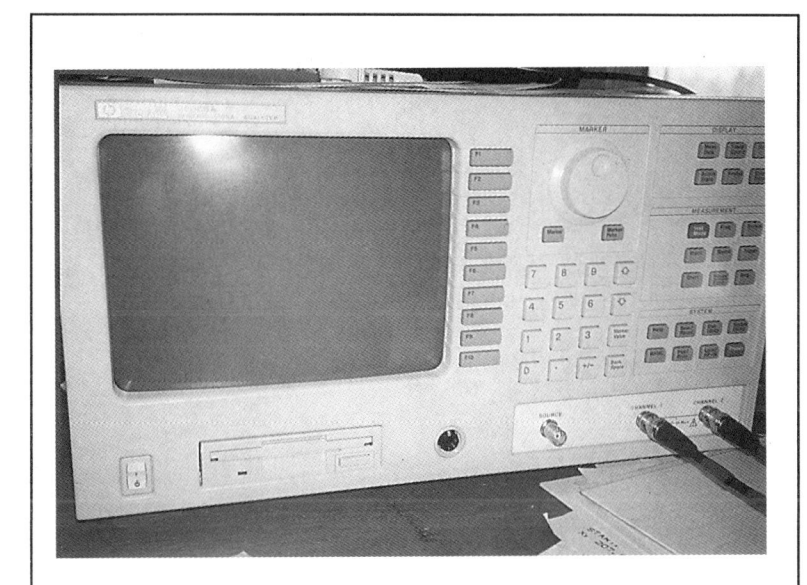

Figure 3.3. Hewlett-Packard FFT Analyzer.
Courtesy of Hewlett-Packard

the screen (see Figure 3.2) and then multiplied by the time base sweep rate (sec/div). When the time base is set on voltage function, it acts in a manner similar to the vertical amplifier, but the signal is a voltage in the horizontal direction that produces an *x-y* display (orbit or lissajous). The screen has ten divisions in the horizontal direction.

FFT Analyzer

The FFT analyzer (Figure 3.3) is a computer-based digital instrument. A block of data digitized in an analog-to-digital converter are processed in a fast Fourier transform (FFT) algorithm to produce a spectrum. The time waveform is reconstructed from the digitized block of data. A dual-channel FFT analyzer allows the properties of and phase between two signals to be obtained. The FFT analyzer has high resolution but can compromise amplitude accuracy, depending on the setup. It is basically an analyzer for taking steady-state rather than transient data.

The analyzer acquires a block of data at a high sampling rate — greater than 200,000 samples per second — depending on the highest frequency span of the analyzer. The analyzer requires that a signal complete an entire cycle before any data can be processed by the FFT. This means that, at low frequencies (below 10 Hz), long sampling times are required before FFT processing begins. The ability of the analyzer to track events when speed is changing rapidly is thus compromised. At common machine frequencies the computational time for FFT processing is a fraction of the data acquisition time and auto ranging time. The zoom function on the FFT analyzer increases resolution; 400 or 800 lines are used, but the frequency span is narrowed to a set frequency (start or center) to achieve frequency resolution. Resolution refers to the capability of the instrument to allow the analyst to view closely-spaced frequencies in the spectrum.

Models 2115 and 2120 Machinery Analyzers.
Courtesy of Computational Systems, Incorporated

DataPAC 1250, DataPAC 1500, and Dataline DSP Data Collectors.
Courtesy of Entek IRD International Corporation

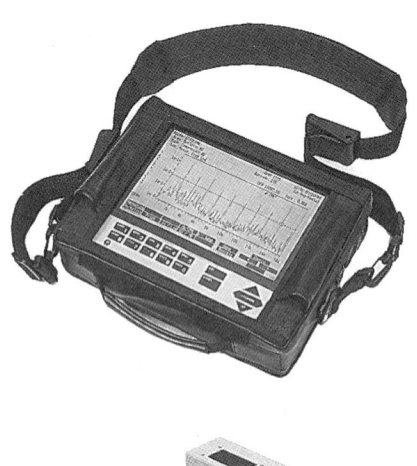

DC-7B and 8603 Data Collectors.
Courtesy of PREDICT/DLI

SpectraVIB Data Collection/Analysis System.
Courtesy of Vibration Specialty Corporation

Figure 3.4. Electronic Data Collectors.

FFT analyzers commonly have more windows available than electronic data collectors. Windows are used to prepare the digitized data for the FFT process. The dynamic range of FFT analyzers is currently close to 72 dB. Thus, a one millivolt signal could be resolved in the presence of a two-volt signal. The many other features of FFT analyzers include orbits, Bodé plots, polar plots, waterfall (cascade) diagrams, and real and imaginary plots used for modal analysis.

Electronic Data Collector

The electronic data collector (Figure 3.4) acquires and stores selected vibration parameters such as overall vibration, overall vibration in selected frequency bands, spectra, time waveforms, orbits, waterfall diagrams, high-frequency measures, and envelope detection spectra. Overall vibration is usually stored as vibration velocity in peak or rms and related to points on a pre-established route, which can include many machines. Data are entered into a computer capable of trending them with previously-collected information, so that any changes in machine condition can be determined.

Electronic data collectors have been sold as analyzers since they were developed. FFT algorithms were introduced into the collectors, and, after several generations, many are now respectable FFT analyzers with adequate resolution and dynamic range. Some collectors are capable of 6,400 lines. Of course the data acquisition time increases by 16 times that of a 400-line spectrum. The number of lines typically available are 100, 200, 400, 800, 1,600, 3,200, and 6,400. The increased number of lines does provide a valid zoom, but expansion is required to view closely-spaced data on the computer screen.

Data Sampling

The input signal from a transducer is digitized prior to FFT processing as shown in Figure 3.5. The number of samples stored in the analyzer buffers depends on the number of lines selected. The computer records these values as equally spaced Y (amplitude) and X (time) components.

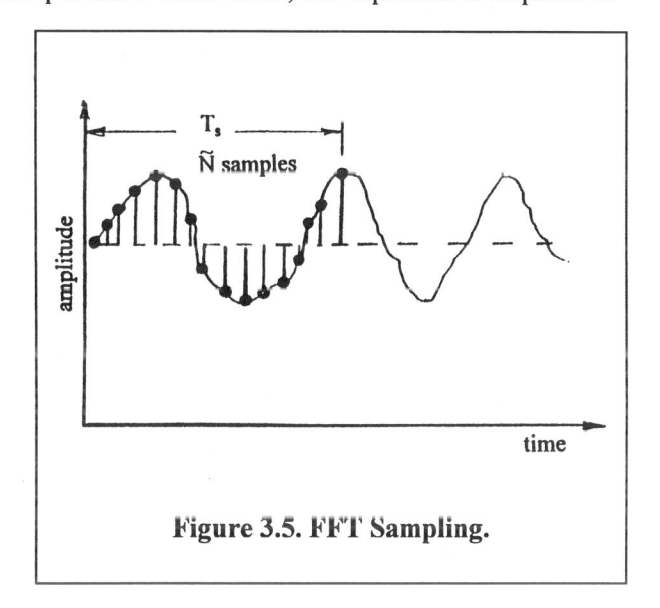

Figure 3.5. FFT Sampling.

Figurre 3.6. Lowest Resolvable Frequency.

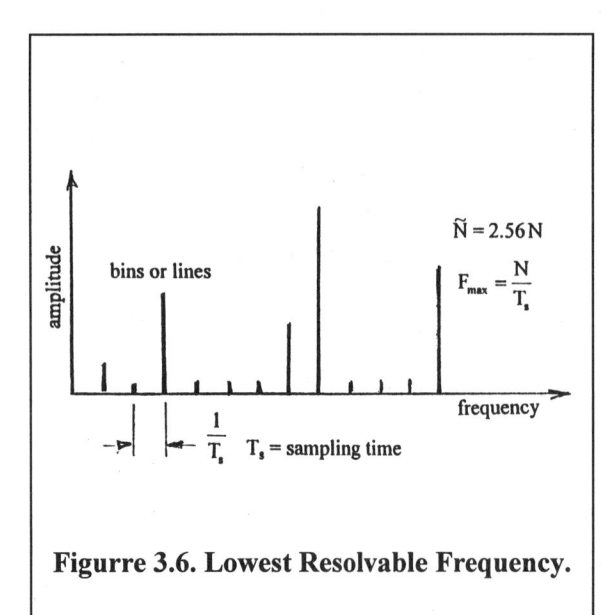

Figure 3.6 shows the equally-spaced samples transformed into a spectrum of N lines or bins. These equally-spaced bins start at the lowest resolvable frequency, which is one divided by the sampling time, or $1/T_s$. No frequency lower than $1/T_s$ can be resolved because the information in the analyzer buffer is incomplete. The spectrum can have N bins (lines) — usually from 100 to 6,400, depending on the number of samples collected by the analyzer or data collector. The number of samples is 1,024 if 400 bins are used. Digital filtering is used to adjust the number of samples acquired over the data acquisition time. The number of samples is thus related to the number of lines selected by a factor of

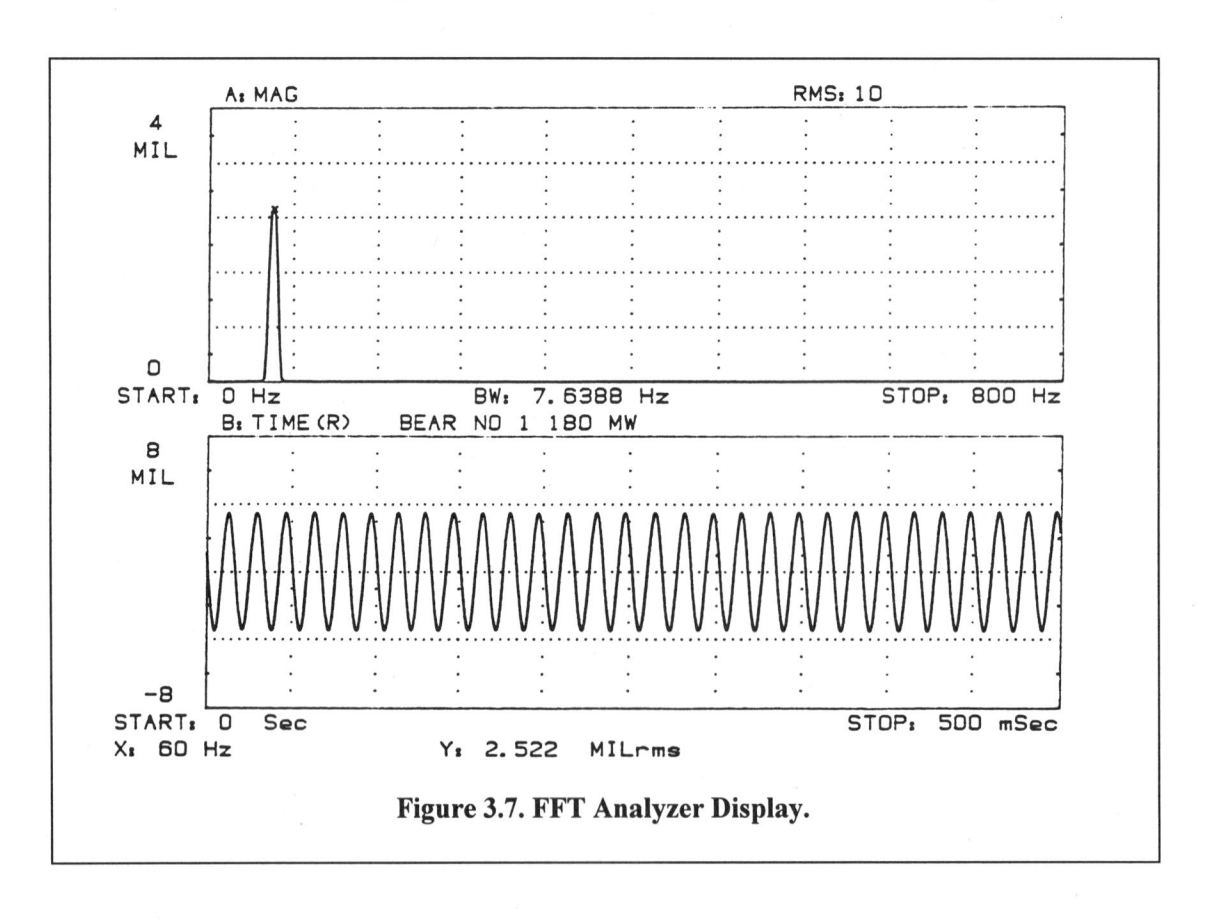

Figure 3.7. FFT Analyzer Display.

2.56. The data sampling rate is adjusted by the analyzer to obtain the number of samples required over the selected data acquisition time (N/F_{max}). The formula for the maximum frequency and data acquisition time is $F_{max} = N/$data acquisition time. The value of F_{max} is set on the analyzer, and the time display automatically has the proper span (Figure 3.7). The F_{max} is 800 Hz; the number of lines is 400. Therefore, the data acquisition time must be $T_s = N/F_{max} = 400/800$, or 0.5 second. This relationship establishes the amount of time required to acquire the data regardless of the speed of the computer.

Aliasing

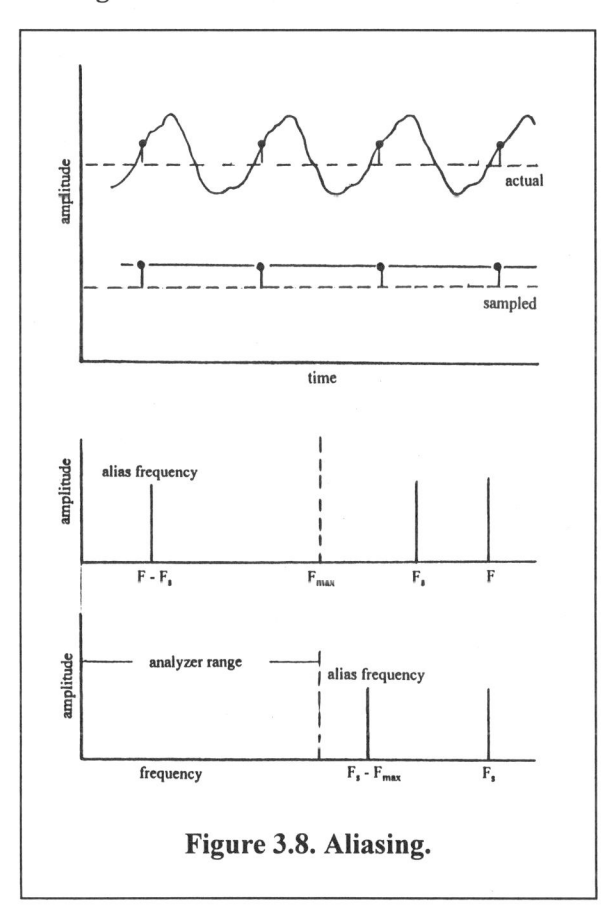

Too infrequent sampling (Figure 3.8) results in lost data and causes false frequencies because of aliasing. This phenomenon occurs in the spectrum if the sampling rate is lower than the frequencies present in the data. Figure 3.8 (top) shows data sampled at the same frequency as the vibration. The resulting digitized data will be a straight line. False or aliased frequencies are obtained in the spectrum when the sampling frequency is lower than the highest frequency present in the data. Therefore, beware of FFT algorithms that do not have anti-aliasing filters. The Nyquist criterion states that the sampling rate of an analyzer must be greater than two times the highest frequency present. Figure 3.9 contains a sampling frequency of two and three times the maximum frequency. The anti-aliasing filter is a low-pass filter that eliminates frequencies in the data that are sufficiently high to cause aliasing (Figure 3.10).

Figure 3.8. Aliasing.

Figure 3.9. Sampling Rates.

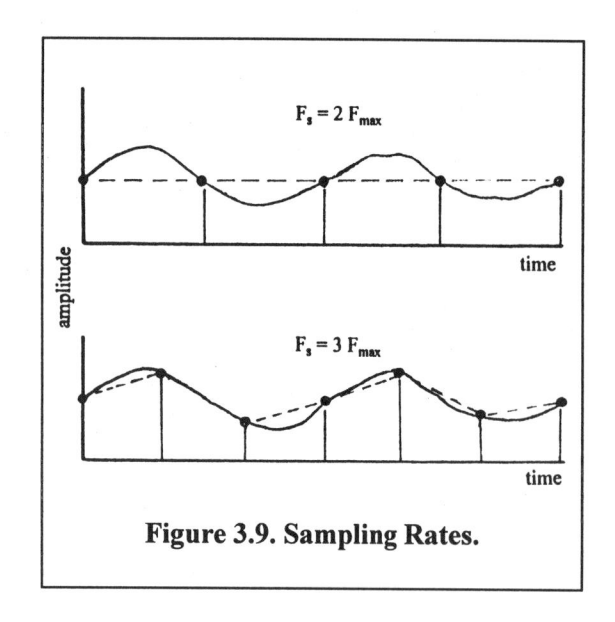

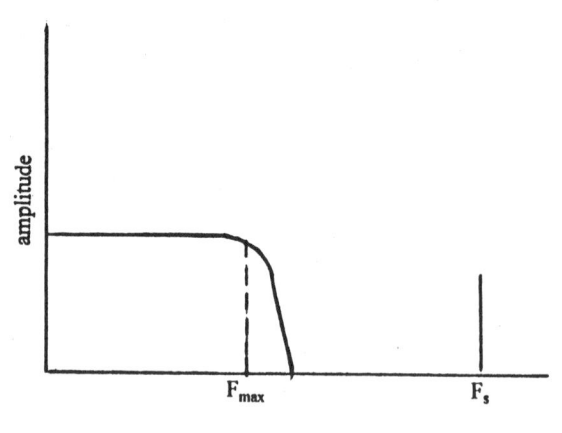

Figure 3.10. Anti-Aliasing Filters.

Windowing

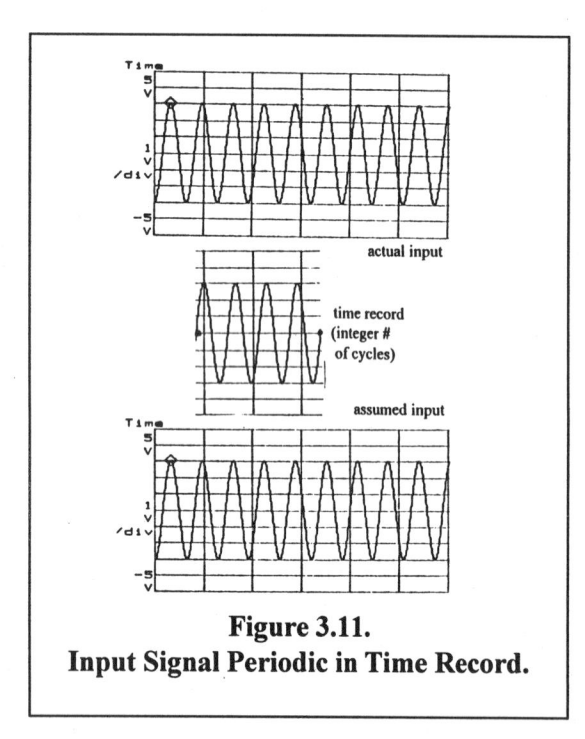

Figure 3.11. Input Signal Periodic in Time Record.

The FFT algorithm that changes the buffer sample — digitized time waveform — into a spectrum assumes that data before and after the sample are similar [3.1]. For this reason, the FFT algorithm requires the sampled data to start and end at zero amplitude (Figure 3.11). Note that the reconstructed waveform is the same as it was prior to sampling. In Figure 3.12 sampling did not occur at zero amplitude, and a reconstructed waveform is obtained that is not the same as the original data. This waveform will cause errors in the spectrum called leakage — that is, energy is spread to higher frequencies. The resulting spectrum, shown in Figure 3.13, implies impact or looseness that causes loss of resolution. Note the wide-banded spectral base around 84 Hz. These data were processed without a window.

Because data acquisition cannot be controlled to obtain periodic sampling, windows are used to force the end points of the data to zero (Figure 3.14). A window function is multiplied by each signal sample to obtain a record that is zero at both ends. Windows need not and should not be used when transient data are taken that start and end at zero. A Hanning window destroys part of the data on a transient signal because valuable data at the beginning of the sampling may be eliminated (Figure 3.15). A uniform or self-windowing function should be used for transient data response. The Hanning window has a narrow filter within the bin or line that allows a bandwidth with good resolution (Figure 3.16). However, because it is narrow within the bin, amplitude inaccuracy (sometimes called bandwidth inaccuracy) is as high as 1.5 dB (18.8%) if the frequency falls at the edge of the bin (Figure 3.17). The flat-top window lacks some resolution,

Table 3.1. FFT Window Selection.

Window	Purpose	Amplitude Uncertainty	Window Factor (WF)
uniform	impact tests	56.5%	1
Hanning	fault analysis	18.8%	1.5
flat top	condition evaluation	1%	3.8

resolution = 2x bandwidth = $\frac{2 \times \text{frequency span} \times (WF)}{\text{number of lines}}$

but it has only a 0.1 dB (1%) amplitude inaccuracy. Therefore, the flat-top window is recommended for discrete spectral lines and amplitude accuracy. The Hanning window is recommended for multi-frequency steady-state data. The actual resolution for each window can be calculated using the window factor provided in Table 3.1. The reliable resolution of the analyzer is two times the frequency span multiplied by the window factor divided by the number of lines selected.

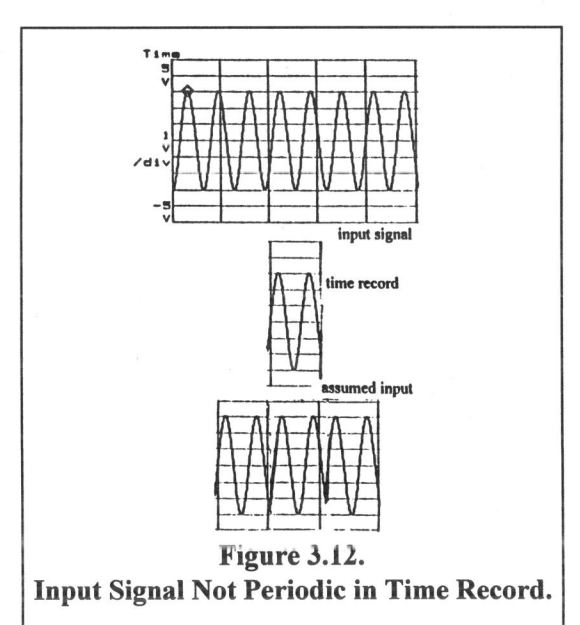

Uniform or rectangular windows are used for responses to impact tests unless the data do not go to zero within the sample. For this reason, double hits within the data acquisition time are not recommended.

Figure 3.12.
Input Signal Not Periodic in Time Record.

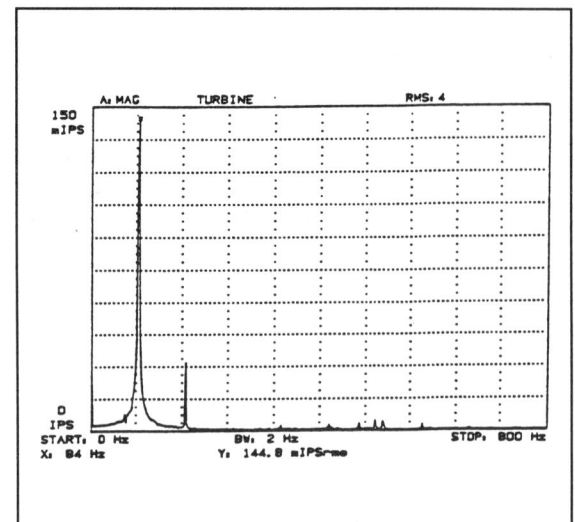

Figure 3.13. Example of Leakage.

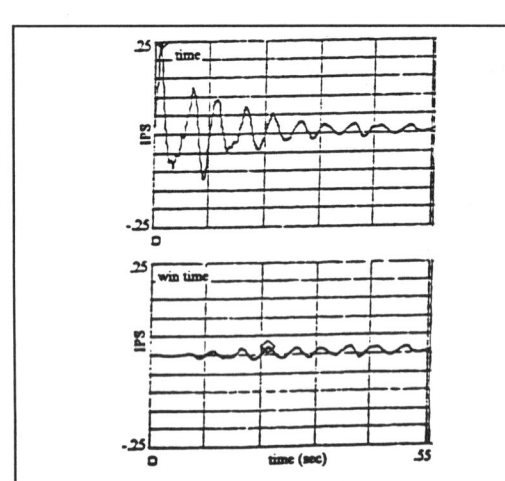

Figure 3.14. The Effect of Windowing on the Time Waveform.

Figure 3.15. Windowing Results in Loss of Information from Transient Events.

Figure 3.16. Resolution of Windows.

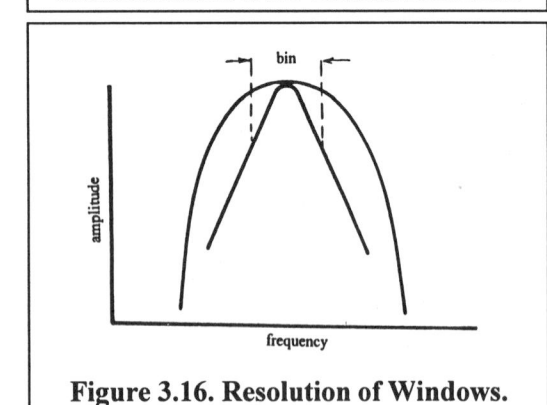

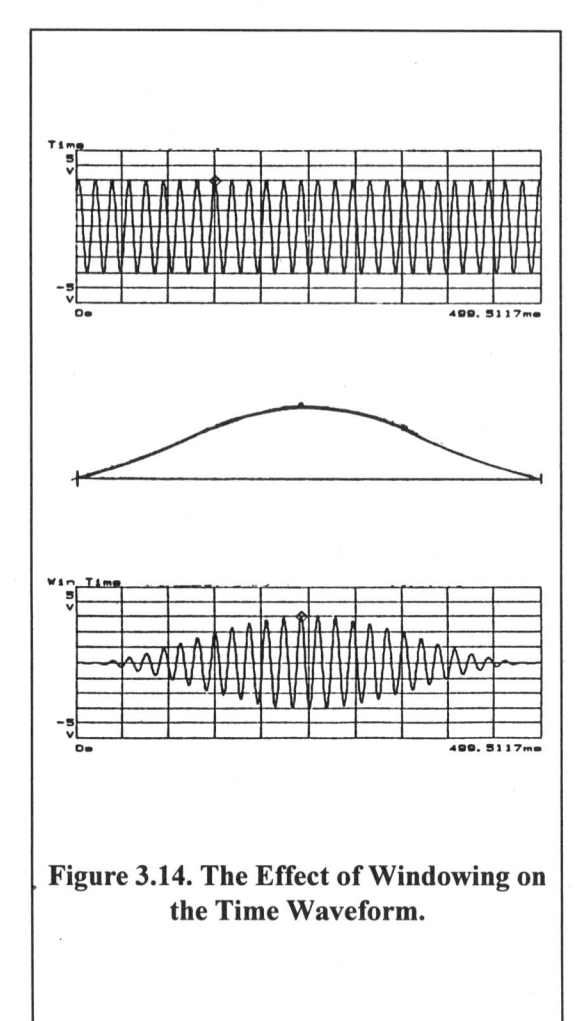

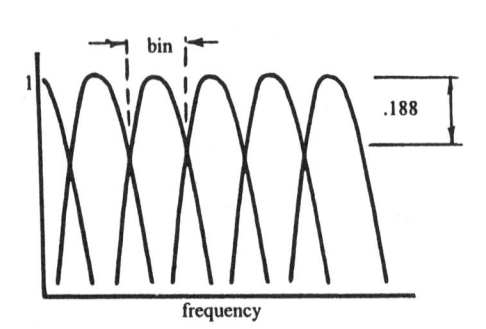

Figure 3.17. Hanning Band-Pass Shapes.

3.10

Dynamic Range

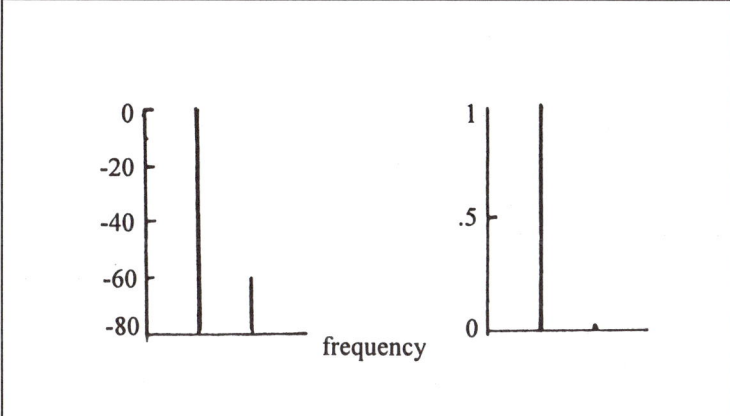

Figure 3.18. Small Signals in the Presence of Large Ones; Second Order Is 1/1000 the First Order.

Dynamic range refers to the capability of an analyzer to resolve small-amplitude components in the presence of high-amplitude components in the spectrum (Figure 3.18). Problems occur in vibration analysis when acceleration or displacement are displayed in a spectrum with a large frequency span, and components are shown at low- and high-frequencies. A wide dynamic range is achieved by using log amplitude scales, which are compressed scales. Figure 3.18 shows both linear and log scales. A second order that is only 0.1% (1/1000) the value of the first order will not appear on the linear scale but will appear at 60 dB less than the first order on the log scale.

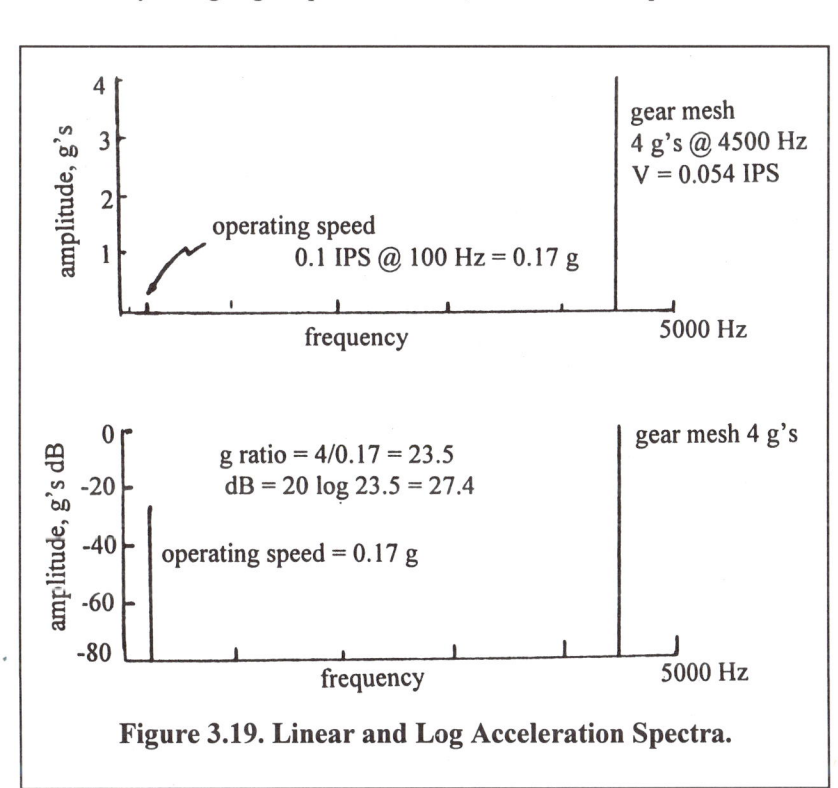

$$dB = 20 \log V/V_{ref}, \text{ or}$$
$$dB = 20 \log 1/1000 = -60$$

This situation may occur in the early stages of rolling element bearing defects and the amplitudes are contained in a spectrum with significant gear vibration amplitudes. An example is 0.7 IPS gear mesh and 0.01 IPS bearing component. The dynamic range required would be

$$dB = 20 \log 0.7/0.01 = 36.9 \text{ dB}$$

Figure 3.19. Linear and Log Acceleration Spectra.

3.11

This dynamic range is available in all modern data collectors; however, the dynamic range must be set up on the analyzer properly or auto ranging should be used.

Good dynamic range is important if low-frequency accelerations or high-frequency displacements are measured in the presence of low-frequency ones. In Figure 3.19, in which gear mesh is seen in the acceleration spectrum as well as a much lower frequency operating speed, 27 dB is required to obtain good dynamic range. The analyzer range should be set so that the signal uses at least half the range span to avoid loss of amplitude resolution.

Averaging

The FFT analyzer can be used in a number of averaging modes other than instantaneous display of the FFT after the data are acquired. These averaging modes include rms, peak hold, synchronous time, and overlap. The rms amplitude average is obtained by averaging the data in the bins — either weighted by sequence of acquisition or not — as each block of data is processed. Noise is smoothed, but not eliminated, and discrete signals are reinforced in the display.

The peak-hold function holds the largest value of the peak or rms in each bin as each new block of data is processed. Therefore, it does not actually do any averaging. The peak hold is used for transient tests. The major disadvantage in using FFT analyzers in transient tests is the long data acquisition time required. A block of data must be processed before it is displayed. Therefore, on a coast-down test, the machine may experience a large change in RPM as a data sample is being collected. Each FFT peak hold yields only one point on the curve. Numerous points are needed to describe the area around the critical speed.

In overlap processing, only a fraction of new data are acquired in the buffer. Data from the previous sample are used to make up a complete sample for processing. The number of lines or bins, frequency span, and overlap processing are considerations when the analyzer is set up for a transient test because they govern data acquisition time.

Synchronous time averaging is performed on the time waveform. A trigger is provided to the analyzer at the shaft frequency from a proximity probe or an optical pickup. The analyzer averages succeeding data sets that are triggered by the shaft rotation. This procedure tends to eliminate signals not synchronous to the trigger and enhances the signal-to-noise ratio of the data. Figure 3.20 and Figure 3.21 show rms and synchronous time-averaged data from a blind drill roll. The synchronous time-averaged data show the vibration related to the blind drill roll.

Figure 3.20. rms-Averaged Vibration of a Blind Drill Roll.

Figure 3.21. Synchronous Time-Averaged Vibration of a Blind Drill Roll.

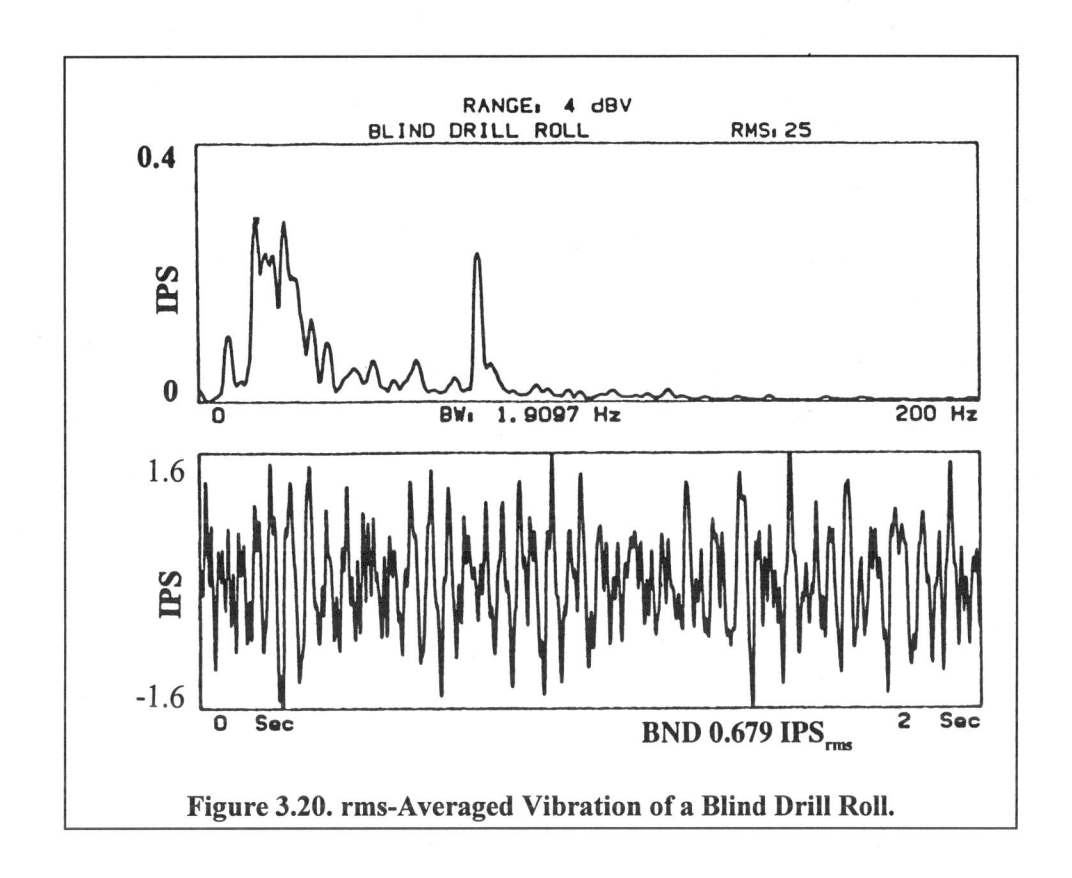

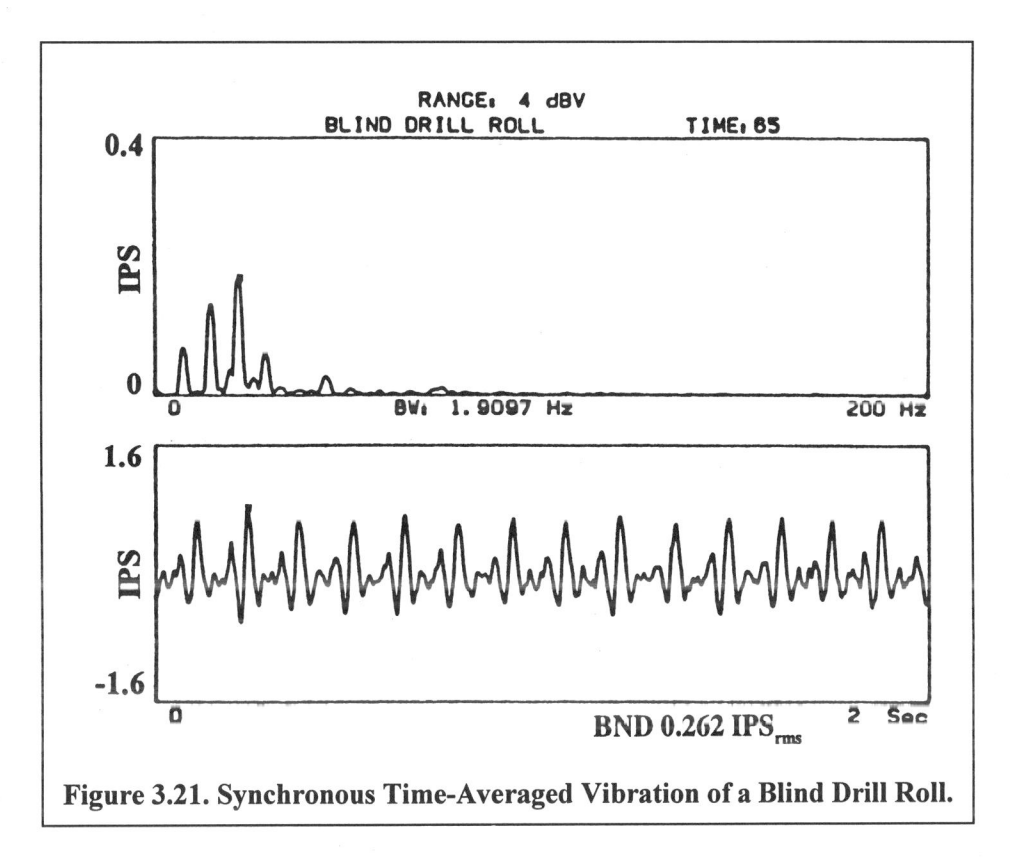

Some FFT analyzers generate a waterfall, or cascade, diagram (Figure 3.22). This diagram contains a number of spectra taken at various speeds or times and, in some cases, space (i.e., the analyzer stacks spectra on a waterfall plot by position of the measurement).

Setup of the FFT Analyzer and Data Collector

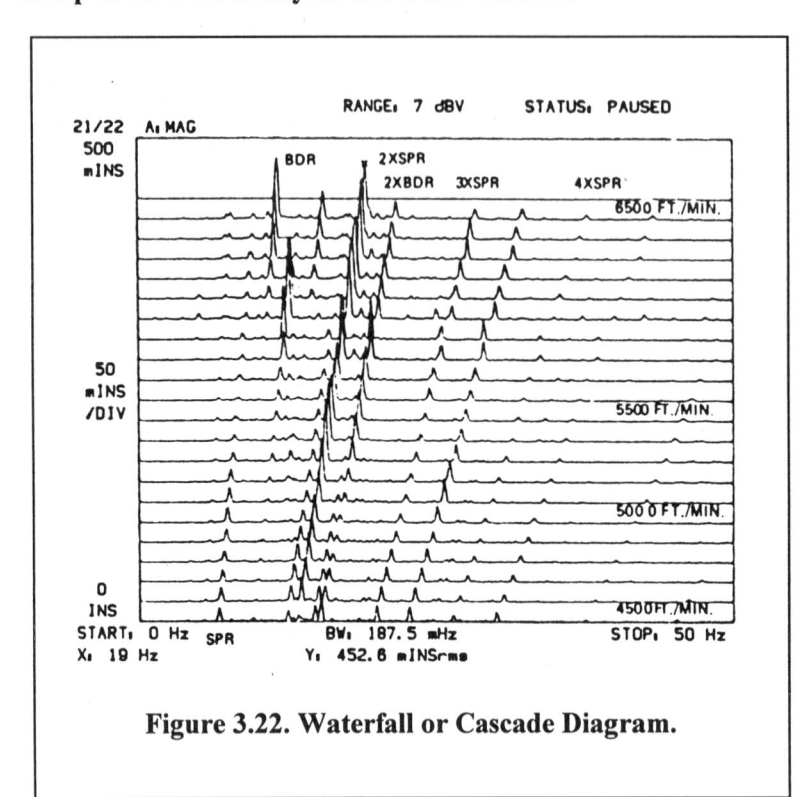

Figure 3.22. Waterfall or Cascade Diagram.

The goal of the FFT analyzer setup is to produce data upon which cost-effective decisions can be made relative to faults and condition. Machine knowledge is essential; i.e., fault frequencies, natural frequencies, and critical fault amplitudes. It is well known that all of these data will not be available when a machine is first monitored or analyzed. However, as time passes and the analyst works with a machine, the experience gained will provide information about the vibration levels at which defects arise that lead to failure. Frequency information with regard to shaft-speed-faults, bearing frequencies, blade- and vane-passing frequencies, and gear-mesh frequencies should be available at the inception of monitoring or analysis. It is therefore possible to do a reasonable setup from machine design data alone. The two principal issues that must be considered are resolution (frequencies) and dynamic range (amplitudes).

Resolution. The resolution present in a spectrum is dependent on the number of lines used in the FFT calculation (which is related to data samples), the frequency span, and the window selected. The data acquisition time for a processed block of data depends on the number of lines and the frequency span. The minimum resolvable frequency is the reciprocal of the data acquisition time. In other words, a full sample (one period) of data at the frequency of interest must be present in the block of data before that frequency is observed in the spectrum. For example, data are to be analyzed at a frequency of an operating speed of 1,800 RPM (30 Hz), then [1/30 =

0.033 sec/cycle]. A minimum of 33 milliseconds of data must be acquired. Otherwise, the vibration at 30 Hz has not made one cycle during the data acquisition process.

Example 3.2: Calculating Lines of Resolution and Data Acquisition Time.

Data are being acquired on a two-pole electric motor with a suspected air-gap problem (vibration is occurring at 120 Hz, or 7,200 CPM). The motor operates at 3,580 RPM. Two times operating speed is 7,160 CPM. The difference between two times line frequency and two times operating speed is 40 CPM. If a frequency span of 500 Hz (30,000 CPM) were chosen, what number of lines of resolution will be required if a Hanning window is used? What will be the data acquisition time?

resolution = 2 x (F_{max}/N) x WF or, rearranging the equation:

no. lines = 2 x (F_{max}/resolution) x WF

no. lines = 2 x (30,000/40) x 1.5 = 2,250

Therefore, 3,200 lines would be required.

data acquisition time = no. lines/frequency span = 3,200/500 Hz = 6.4 seconds

---

No frequencies can be resolved between lines. This fact is important in relating frequency span and number of lines to the lowest resolvable frequency. If 400 lines are selected, the spectrum will be divided into 400 discrete points, with all frequencies falling between the lines loaded into their adjacent line. The analyzer displays the frequency at the center of the bin. For example, if a frequency span of 1,000 Hz had been selected in the example, the lowest resolvable frequency would be 1,000 Hz/400 lines, or 2.5 Hz. Therefore, 400 milliseconds (1/2.5 = 0.40 second) of data would have been acquired, and 30 Hz could be resolved.

Error and noise are introduced when windows are utilized in FFT processing. For this reason a window factor (noise factor) is used to calculate guaranteed resolution. The theoretical resolution of frequency span divided by the number of lines must thus be downgraded by multiplying by two times the window factor (Example 3.2).

When a data collector is used to conduct fault diagnosis or condition evaluation, a time waveform should be stored with each spectrum. If the data are being tape recorded, the data stored should be sufficient so that FFT analysis can be conducted to the desired frequency spans and resolution. The reliable resolution (2x bandwidth) is

resolution = [2 x (frequency span) x (window factor)]/no. lines

The data acquisition time remains the number of lines divided by the frequency span. Examples 3.2, 3.3, and 3.4 are concerned with setting up the data collector.

Example 3.3: Lines of Resolution.

Data are being acquired from a fan operating at 956 RPM. The fan is mounted close to a second fan that operates at 970 RPM. How many lines of resolution are required to resolve shaft frequency using a Hanning window and a minimum frequency span of 10 times the operating speed (9,700 CPM)?

Since 12,000 CPM is an available frequency span on the data collector, it is selected.

$$\text{resolution required} = 970 \text{ CPM} - 956 \text{ CPM} = 14 \text{ CPM}$$

$$\text{Then } 14 \text{ CPM} = [2 \times (12,000 \text{ CPM}) \times 1.5]/\text{no. lines}$$

$$\text{or no. lines} = [(2 \times 12,000 \text{ CPM}) \times 1.5]/14 \text{ CPM} = 2,571 \text{ lines}$$

Therefore, 3,200 lines, the next highest number of lines available on the data collector, would provide sufficient resolution.

Example 3.4: Data Acquisition Time.

What is the data acquisition time for the FFT setup used in Example 3.3 if ten averages with 25% overlap are used?

$$\text{data acquisition time} = 3,200 \text{ lines}/[12,000 \text{ CPM}/60] = 3,200/200$$

$$= 16 \text{ seconds (first sample)}$$

$$\text{data acquisition time} = (16 \text{ seconds})(1.0 - \text{overlap}) = \text{second to tenth sampling}$$

$$\text{total data acquisition time} = 16 \text{ seconds} + [(0.75)(16)(9)] = 124 \text{ seconds}$$

Dynamic range. Dynamic range determines whether or not amplitudes at different frequencies can be resolved. Most data collectors (12 bit) have 72 dB of dynamic range; however, one bit is used for the +/- sign (thus, $2^{11} = 2,048$). This means that AC signals with a difference in amplitude up to 2,000 to 1 can be resolved (DC signals of 4,000 to 1). Older 8-bit data collectors have a dynamic range of 42 dB (128-1). If the input range of the FFT analyzer is set too high for the difference in amplitudes of two signals, the dynamic range will not be sufficient to allow the lower amplitude signal to be observed.

The best dynamic range is obtained when the range is set as close to the highest signal strength possible. Auto ranging is commonly used. For example, if a steady velocity is being measured at 0.45 IPS with a 12-bit data collector, the range should be set at 1 IPS, not 5 IPS. The reason is that, at 1 IPS almost full dynamic range can be obtained. Signals as low as 0.0005 IPS (1.00/2,000) can be detected. At a setting of 5 IPS, however, signals only as low as 0.0025 IPS (5/2,000) would be detected.

Auto ranging and auto scaling have been made features of data collectors to assure that the best dynamic range and vertical scales are selected to acquire data without overloading the collector. Overloading causes clipped signals that are useless in processing because false orders and harmonics are generated by the FFT process. A negative effect of auto ranging is lost time. If the signal has wide variations in amplitude, a minimum of one block of data is necessary to auto range.

Summary of Data Processing
- The oscilloscope measures voltages that vary with time and provides a time waveform.
- The voltage, peak or peak to peak, in the time waveform can be converted to a peak-to-peak displacement, peak velocity (largest value), or peak acceleration. The voltage must be divided by the transducer scale value; for example, 100 mv per g for an accelerometer.
- Vibration frequencies can be calculated from the period (time of repetition) of the time waveform and inverting it.
- The oscilloscope can display orbits (*x-y*) of shaft or pedestal motion by making the time base a horizontal amplifier.
- Shaft rotational speed and number of revolutions of the shaft per cycle of vibration can be displayed using the *z*-axis intensification of the oscilloscope.
- Triggering signals obtained from a proximity probe or an optical pickup can be used to initiate and/or control data acquisition.
- The FFT analyzer displays a time waveform and a spectrum from a digitized block of data.
- FFT data are displayed in discrete lines (bins). Frequencies between these lines cannot be distinguished.
- The lowest resolvable frequency is one over the data acquisition time.
- The data acquisition time in seconds for a block of data used by an FFT analyzer to produce a spectrum is equal to the number of lines divided by the frequency span in Hertz.

- Aliasing — that is, too infrequent sampling — causes false frequencies to be displayed in the spectrum.
- Data are sampled at least two times the highest data frequency to avoid aliasing.
- The lack of absolute periodicity in vibration results in leakage (harmonics) in the spectrum.
- Windows are used in the FFT analyzer to force a block of data to start and end at zero.
- The Hanning window is used in general data collection because it is a good compromise between amplitude accuracy and frequency resolution.
- The uniform window is used for impact testing because it starts and ends at zero without compromising the initial data in the sample.
- The amount of resolution determines whether or not closely-spaced frequencies can be viewed on the spectrum.
- The resolution of an FFT analyzer setup is calculated by multiplying two times the window factor by the ratio of the frequency span to the number of lines.
- Dynamic range is concerned with the amplitude of the spectrum.
- The dynamic range of an FFT analyzer allows small amplitudes to be resolved in the presence of large ones.
- Averaging is used to enhance the data in the time waveform and the spectrum.
- Noise in the spectrum is statistically averaged, not eliminated, by rms averaging.
- Synchronous time averaging in the time waveform eliminates all noise and vibration not related to the trigger frequency.
- Peak hold is used in the FFT analyzer to hold the largest value of the data processed in each bin.
- A waterfall diagram is a three-dimensional plot of spectra versus speed, time, or space.

Reference

The Fundamentals of Signal Analysis — Application Note 243, Hewlett-Packard, 1501 Page Mill Road, Palo Alto, CA 94304 (June 1982).

CHAPTER IV
FAULT DIAGNOSIS

Frequencies are the key to the analysis.

The frequencies measured on the bearing caps and shafts of a machine are used to conduct fault diagnosis. These vibrations are caused (excited) by vibratory forces (excitations). In general, the frequency of the measured vibration is the same as that of the force causing the vibration. The forces arise from machine wear, installation, and design faults. Impulsive forces sometimes excite natural frequencies, which are properties of the system and typically do not change directly with operating speed. However, in machines with fluid-film bearings, natural frequencies may be altered by operating speed.

The ease with which a fault can be identified from good test data is directly proportional to the information available about the design of the machine and its working mechanisms, especially when the same frequencies can be used to identify different faults; e.g., mass unbalance, looseness, and misalignment. The operating speed of the machine is the reference frequency for diagnostic techniques. Other frequencies are either related to operating speed or are shown to be unrelated. A multiple of operating speed (order) implies that a vibration arises as a result of machine operation. Other frequencies, such as those obtained from rolling element bearing defects, are not orders of operating speed but are nonsynchronously related; that is, they are some fraction of operating speed.

Fault Diagnosis Techniques

The basic techniques (Table 4.1) used to conduct fault diagnosis utilize the time waveform, orbit, spectrum, and phase. The frequencies that are acquired from shaft displacement and casing-mounted transducers are related to the known frequencies of the machine. The shape and frequency of the time waveform and the orbit provide insight into the physical characteristics of the motions of the shaft and casing. Phase shows the time relationship between vibrations measured at various locations on the machine; this is called relative phase. Phase also provides information about the time relationship between vibration at one machine location and a fixed reference on the shaft or casing; this is termed absolute phase. The spectrum is an amplitude versus frequency record of the vibration activity at a specific location on the machine.

Table 4.1. Diagnostic Techniques for Rotating Machinery.

Technique	Use	Description	Instrument
time waveform analysis	modulation, pulses, phase, truncation, glitch	amplitude vs time	analog and digital oscilloscope, FFT analyzer
orbital analysis	shaft motion, subsynchronous whirl	relative displacement of rotor bearing in *XY* direction	digital vector filter, oscilloscope
phase analysis	force/motion relationships, vibration/space relationships	relative time between force and vibration signals or between two or more vibration signals	strobe light, digital vector filter, analog and digital oscilloscope, FFT analyzer
spectrum analysis	direct frequencies, natural frequencies, sidebands, beats, subharmonics, sum and difference frequencies	amplitude vs frequency	FFT analyzer, electronic data collector

Spectrum analysis. A spectrum can be analyzed quickly using the following steps.

- Identify operating speed(s) and its multiples (orders). The data in Figure 4.1 are in a frequency and order format.

- Identify dominant frequencies that are multiples of operating speed. Included are blade pass in fans, vane pass in pumps, and gear mesh in gears (Figure 4.2).

- Identify nonsynchronous multiples of operating speed such as bearing frequencies (Figure 4.3).

- Identify beat frequencies — two frequency components close to each other; their amplitudes add and subtract during the beat cycle (Figure 4.4).

- Identify frequencies that do not depend directly on operating speed such as natural frequencies or frequencies from adjacent machines (Figure 4.5).

- Identify sidebands (Figure 4.6) that are related to a low-frequency component of vibration that modulates (changes) the amplitude of a high-frequency vibration. Sidebands are frequency components that appear in the spectrum in addition to a dominant frequency such as gear mesh. Modification of the gear-mesh vibration of a gearbox by uneven wear (see Figure 4.6) is a good example. A sideband identifies the location of a fault if its frequency matches the speed of a machine component.

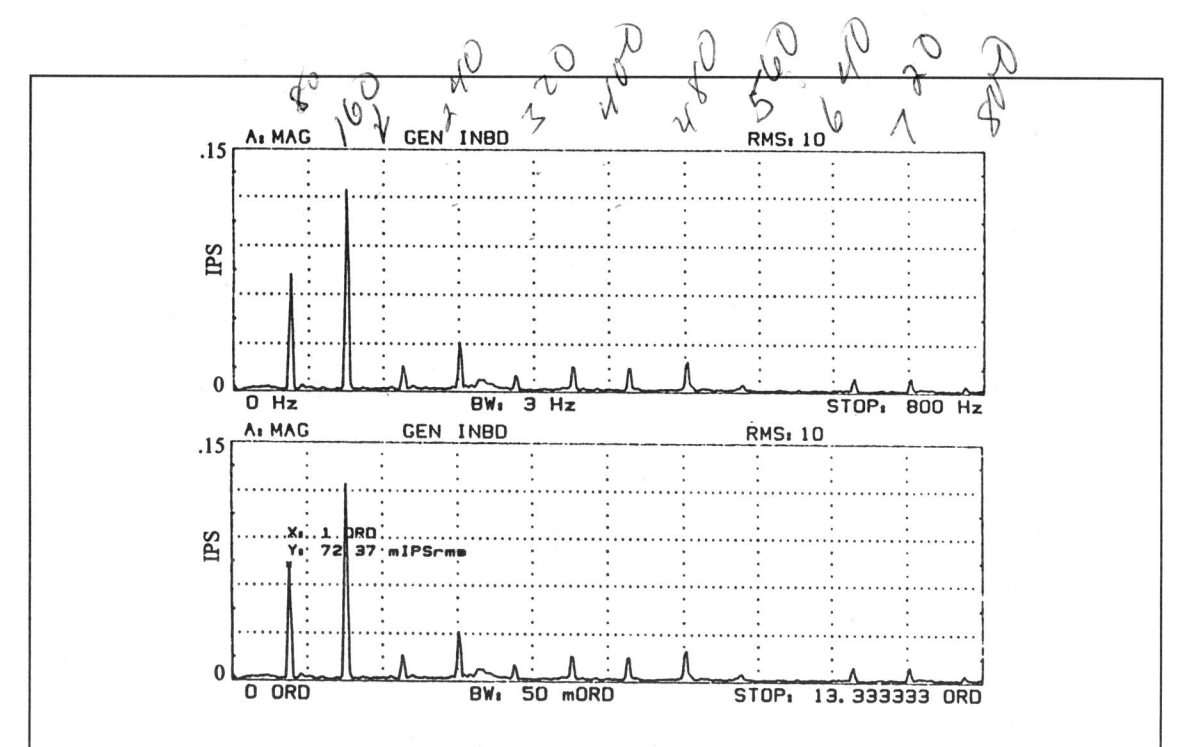

Figure 4.1. Air-Gap Vibration Induced in a Turbine Generator by Misalignment. The Spectrum Shows Order and Frequency.

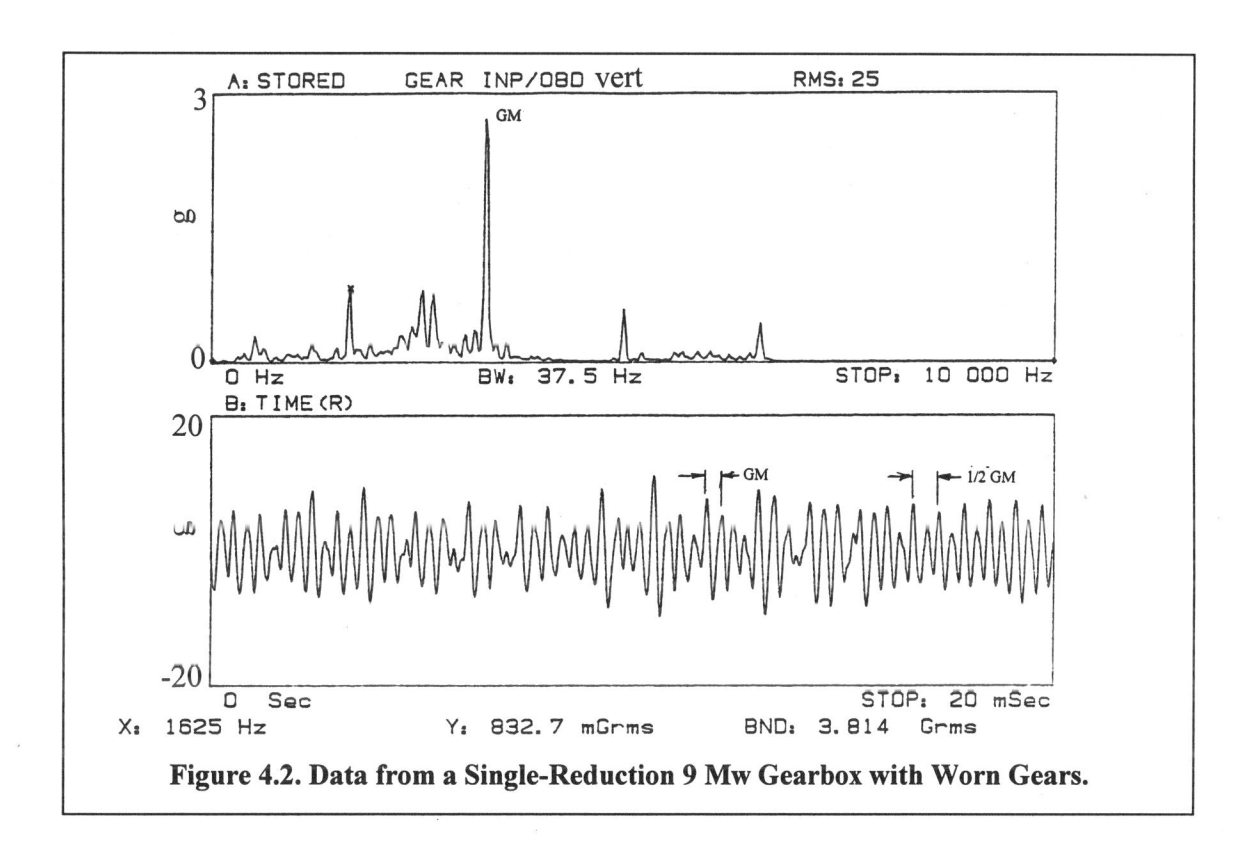

Figure 4.2. Data from a Single-Reduction 9 Mw Gearbox with Worn Gears.

4.3

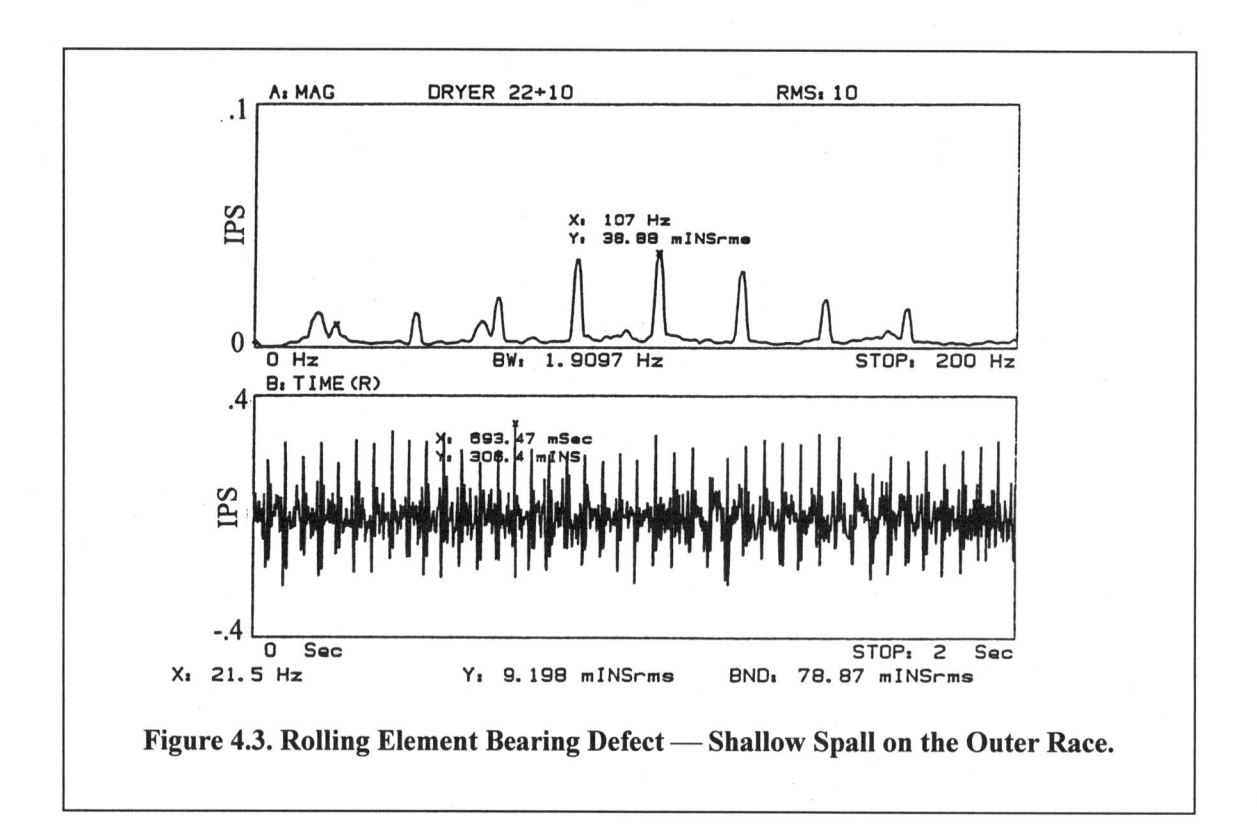

Figure 4.3. Rolling Element Bearing Defect — Shallow Spall on the Outer Race.

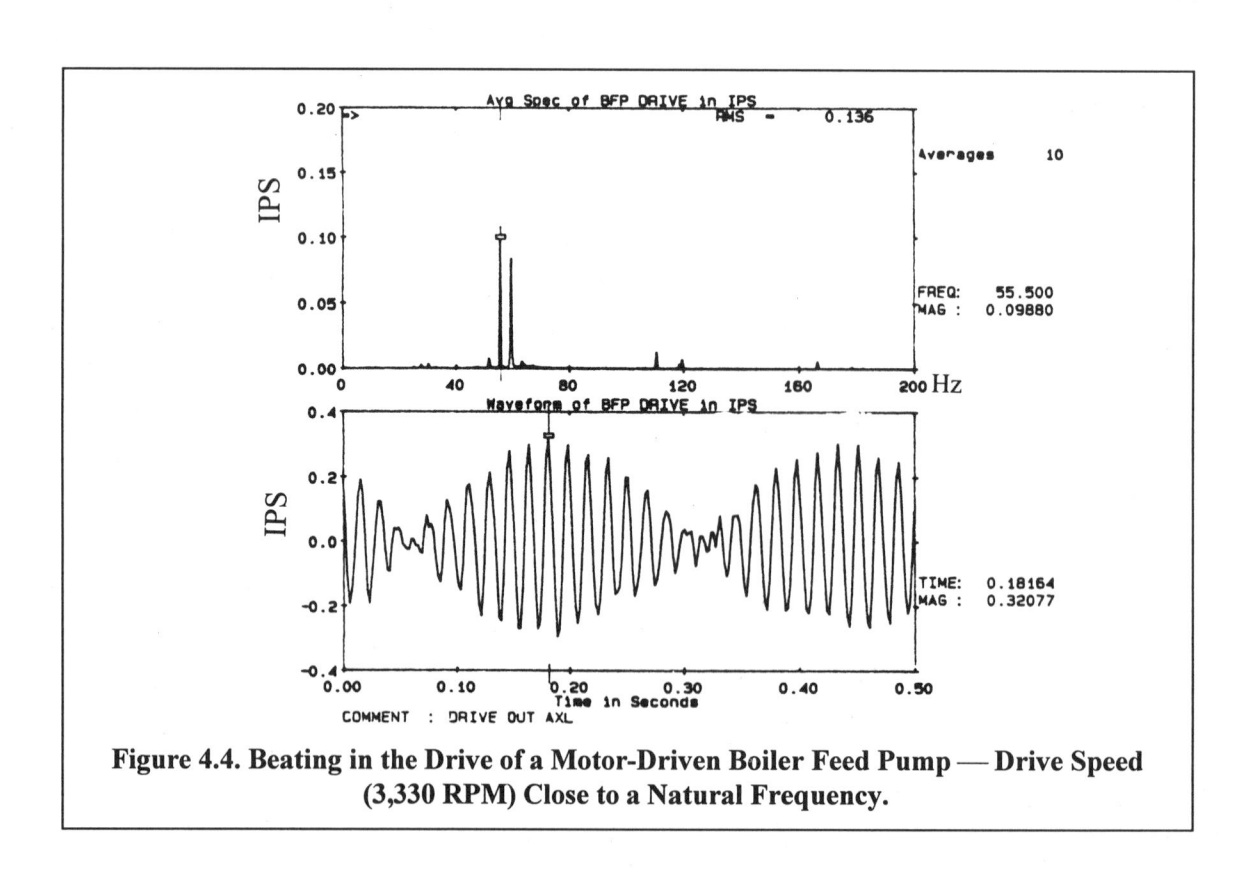

Figure 4.4. Beating in the Drive of a Motor-Driven Boiler Feed Pump — Drive Speed (3,330 RPM) Close to a Natural Frequency.

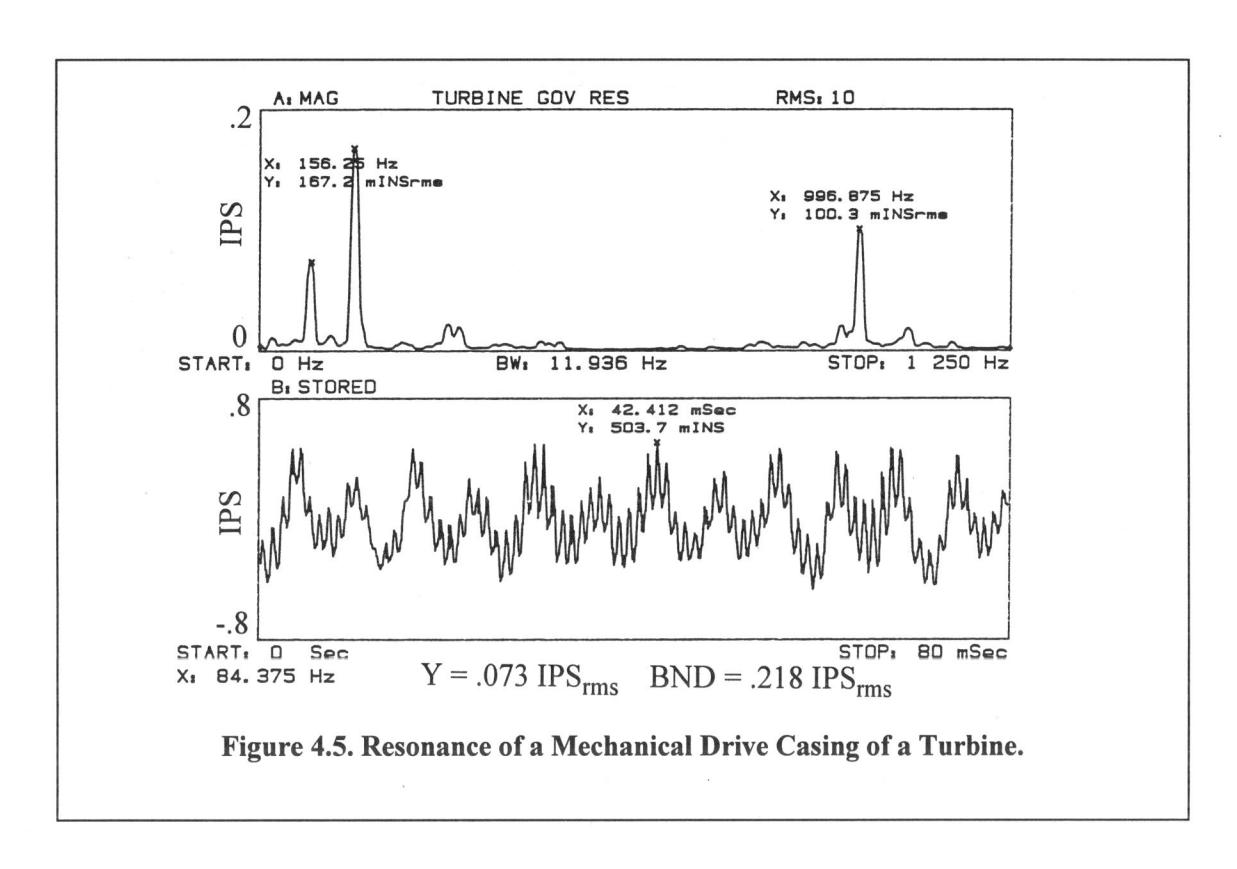

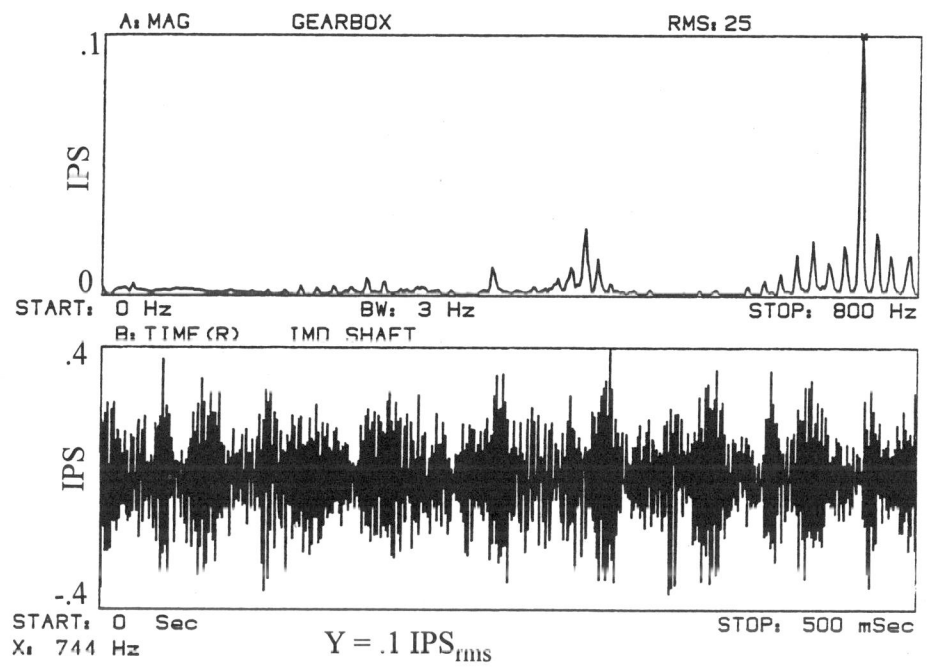

Figure 4.5. Resonance of a Mechanical Drive Casing of a Turbine.

Figure 4.6. Double-Reduction Gearbox – Right-Angle Drive with Improper End Float.

4.5

Spectral identification. It is a fact of life that each instrumentation company uses a different format for presenting data. Therefore, the nomenclature for formats similar to Figure 4.5 is reviewed.

- **A:MAG** — the vertical axis of the spectrum in root mean square (rms) units. Peak is calculated by multiplying the amplitude of the spectral component by 1.414.
- **rms:10** — ten is the number of rms averages in the spectrum.
- **IPS** — inches per second.
- **STOP:1250 Hz** — the f_{max} is 1,250 Hz or 75,000 CPM.
- **B:STORED** — the time waveform. It is typically indicated by B: TIME.
- **BW:11.936 Hz** — the bandwidth.
  (F_{max}/no. lines) (window factor). This analyzer has a fixed number of lines (400). A flat top window was used when the data were acquired. Therefore, BW = (1250 Hz/400 lines) (3.82) 11.94 Hz.
- **STOP:80 msec** — the time waveform has an 80 msec span, or 0.0080 second per division.
- **X:84.375 Hz** — horizontal coordinate (frequency in Hz) in the spectrum at the cursor marker.
- **Y:0.073 IPS$_{rms}$** — vertical coordinate (amplitude in rms) at the cursor marker. 0.073 IPS$_{rms}$ = 0.104 IPS peak.
- **BND:.218 IPS$_{rms}$** — overall rms value of components in the spectrum over the frequency span selected.
- **RANGE:4 dBv** — dynamic range in volts.

Operating Speed Faults

Operating speed faults occur at a predominant frequency of operating speed and its orders (multiples of operating speed). Table 4.2 lists some of the malfunctions that can be associated with machine speed.

Critical speeds. An excitation with a frequency close to or equal to a natural frequency under conditions of low damping (less than 15% of critical damping) is defined as a resonance. If the resonance is caused by a rotating machine, it is termed a critical speed. The interference diagram (Figure 4.7) illustrates the concept of an excitation equal to a natural frequency at various rotor speeds. The horizontal axis is a plot of rotor speed in RPM. The units of vibration frequency in the vertical axis are in CPM. Frequencies of the machine forces, that is, forcing frequencies, are

Table 4.2. Identification and Correction of Malfunctions of Rotating Machinery — Operating Speed Defects.

Fault	Frequency	Figure #	Spectrum, Time Waveform, Orbit Shape	Correction
critical speeds	1x, 2x, 3x, etc.	4.7	amplified vibration due to proximity of operating speed to natural frequency	tune natural frequency
mass unbalance	1x	4.8	distinct 1x with much lower values of 2x, 3x, etc.; elliptical and circular orbits; constant phase	field or shop balancing
misalignment	1x, 2x, occasionally 3x	4.9, 4.10	distinct 1x with equal or higher values of 2x, 3x; 1x axial	perform hot and/or cold alignment
shaft bow	1x	4.11	dropout of vibration around critical speed in Bodé plot	heating or peening to straighten rotor (allow rotor to float axially)
fluid-film bearing wear and excessive clearance	1x, sub-harmonics, orders	4.12	high 1x, high 1/2 x, sometimes 1-1/2 or orders; cannot be balanced	replace bearing
resonance	1x, 2x, 3x, etc.	4.5	high balance sensitivity, high-amplitude vibration at order of operating speed	change structural natural frequency
looseness	1x plus large number of orders, 1/2x may show up	4.13	high 1x with lower-level orders, large 1/2 order, low axial vibration	shim and tighten bolts to obtain rigidity
eccentricity	1x	4.14	high 1x	machine journal for concentricity
coupling lockup	1x, 2x, 3x, etc.	4.10	1x with high 2x similar to misalignment; start and stops may yield different vibration patterns	replace coupling or remove sludge
thermal variability	1x		1x has varying phase angle and amplitude with load	compromise balance or remove problem
distortion	1x and orders		1x from preload of bearings, 2x line frequency, air gap on motor	relieve soft foot

4.7

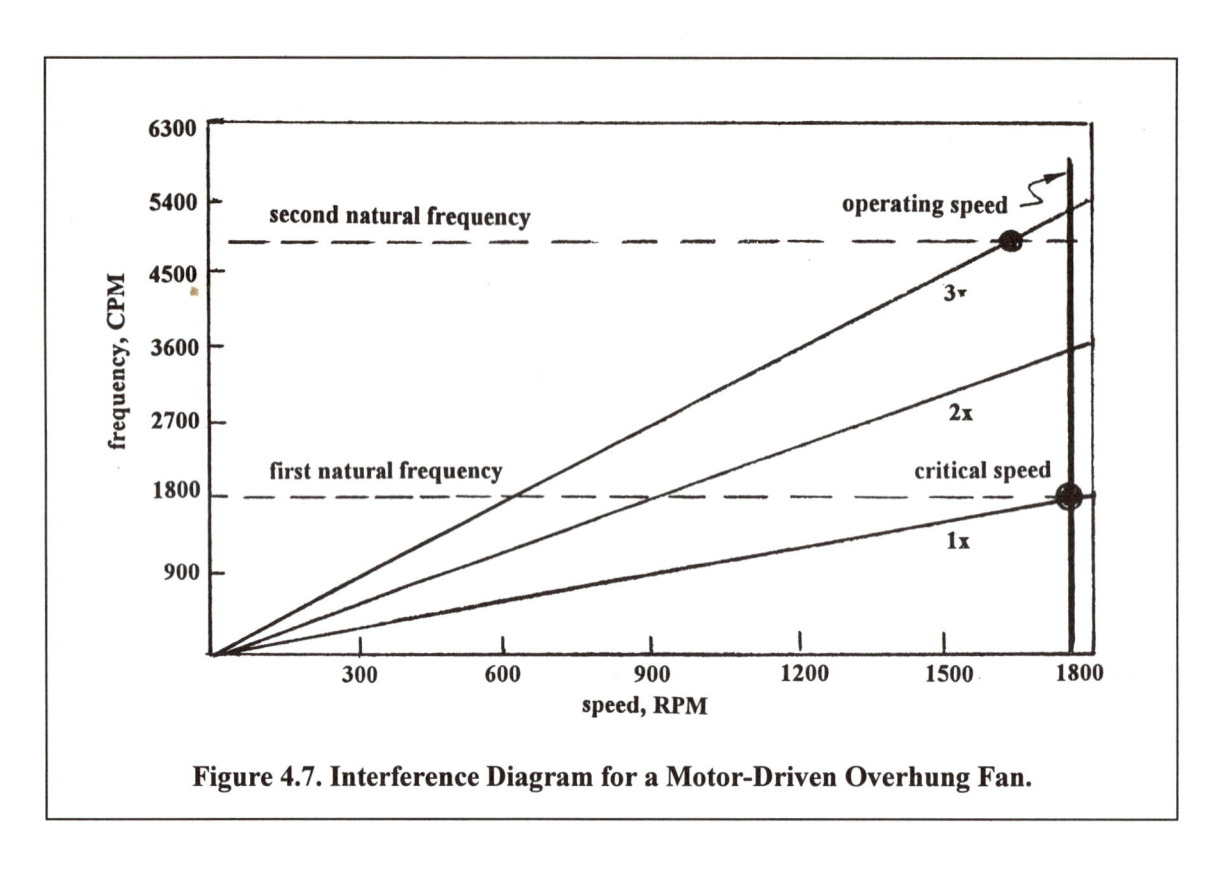

Figure 4.7. Interference Diagram for a Motor-Driven Overhung Fan.

plotted against rotor speed. The vibration from components of operating speed vibration will increase in proportion to the machine speed. For example, at 1,800 RPM the vibration at operating speed (1x) is 1,800 CPM (30 Hz). The frequency of the component at two times operating speed (2x) is 3,600 CPM (60 Hz). The line increases at two times the rate of the vibration at operating speed (1x). The natural frequencies in the range of interest are plotted against operating speed and are usually calculated design curves. The machine is typically tested only at critical speeds (see *Critical Speed Measurements; Chapter VI*).

Critical speeds are excited by rotor speed or its multiples (orders) when a natural frequency and a forcing frequency are equal. Operation of a machine close to a natural frequency amplifies vibration, depending on the damping present and the proximity of the operating speed to the natural frequency. A spectrum of a machine operating close to the natural frequency of a rotor bearing is shown in Figure 4.4. If the operating speed is higher than one or more critical speeds, the rotor must be able to pass through them with acceptable vibration levels; that is, the machine must be well balanced, and some damping is necessary in the bearing. It is true that rotors can be driven through their critical speeds if sufficient power is applied and vibration response is attenuated by a fast start-up. However, the rotor coasts down at a rate that is dependent on its inertia and any frictional or aerodynamic forces present. No control is possible. A rotor can therefore remain at a critical speed for a dangerously long time during coast down, especially if it is operating in a low-density fluid. This situation occurs when a steam turbine is shut down. It is not good to operate a machine at a critical speed because the vibration is amplified (resonant behavior).

Mass unbalance. Mass unbalance occurs when the geometric center (shaft centerline) and the mass center of a rotor do not coincide. Unbalance is a once-per-revolution fault — that is, it occurs at the frequency of rotor speed (Figure 4.8) — and is sometimes difficult to distinguish from misalignment. However, unbalance causes a rotating force; the force of misalignment is

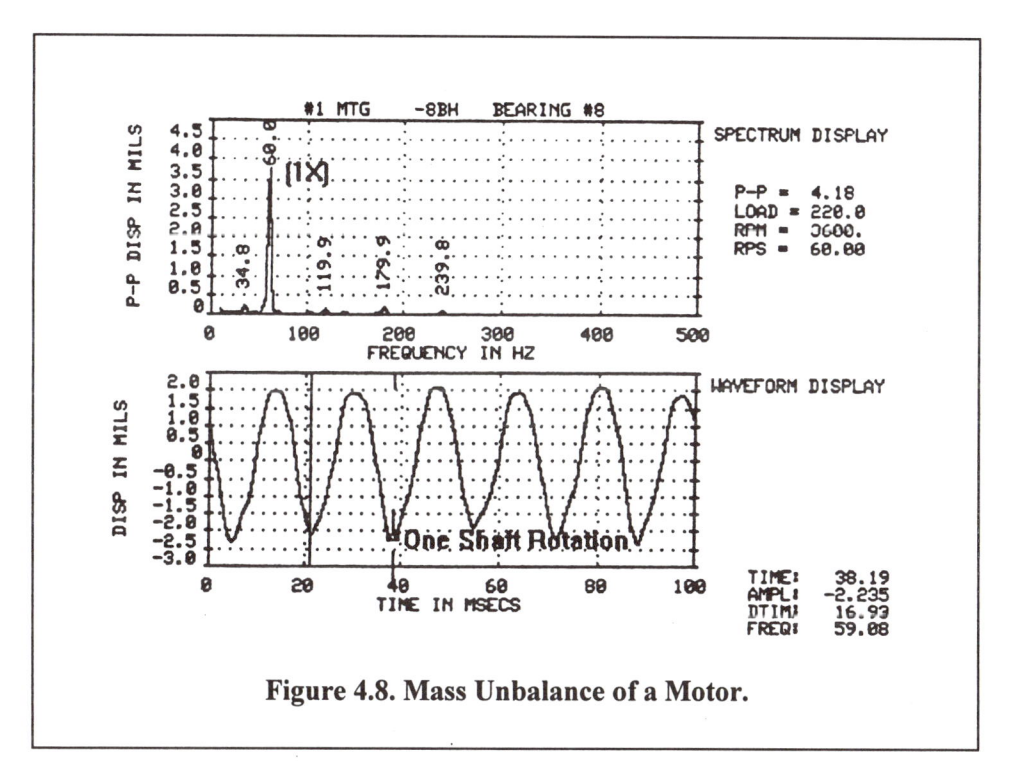

Figure 4.8. Mass Unbalance of a Motor.

directional. Mass unbalance has a fixed phase angle with respect to a reference mark on the shaft. The spectrum has low-amplitude higher-order frequency components. Unlike normal conditions when motions are sinusoidal, nonlinear behavior of a bearing or pedestal in the presence of excessive mass unbalance can cause truncated motions that introduce higher-order vibrations (e.g., 2x, 3x) with amplitudes less than operating speed. Because of the design of the machine the horizontal amplitudes will usually be larger than those measured in the vertical direction. Another factor that affects the relative magnitude of the horizontal-to-vertical components is the proximity of the operating speed or one of its orders to a natural frequency (see Figure 4.4). The axial component of vibration is usually small. However, a large component at operating speed as a result of mass unbalance can cause pedestal motions in the axial bearing, depending on machine design.

Misalignment. Misalignment in a redundantly-supported rotor — that is, a rotor with three or more radially-loaded bearings — causes a rotating preload in the bearings, shaft, and external couplings at the frequency of shaft speed. The magnitude of the resulting vibration is dependent on the radial stiffness of the components (bearings, shafts, seals, couplings) in the system. Severe misalignment can cause nonlinear bearing behavior in one or both directions, depending on asymmetry in the bearing and the stiffness of the pedestal and foundation. The nonlinear behavior causes clipped (truncated) waveforms and/or nonlinearly-generated second and higher orders of shaft vibration (Figure 4.9). The second order component of vibration in cases of severe

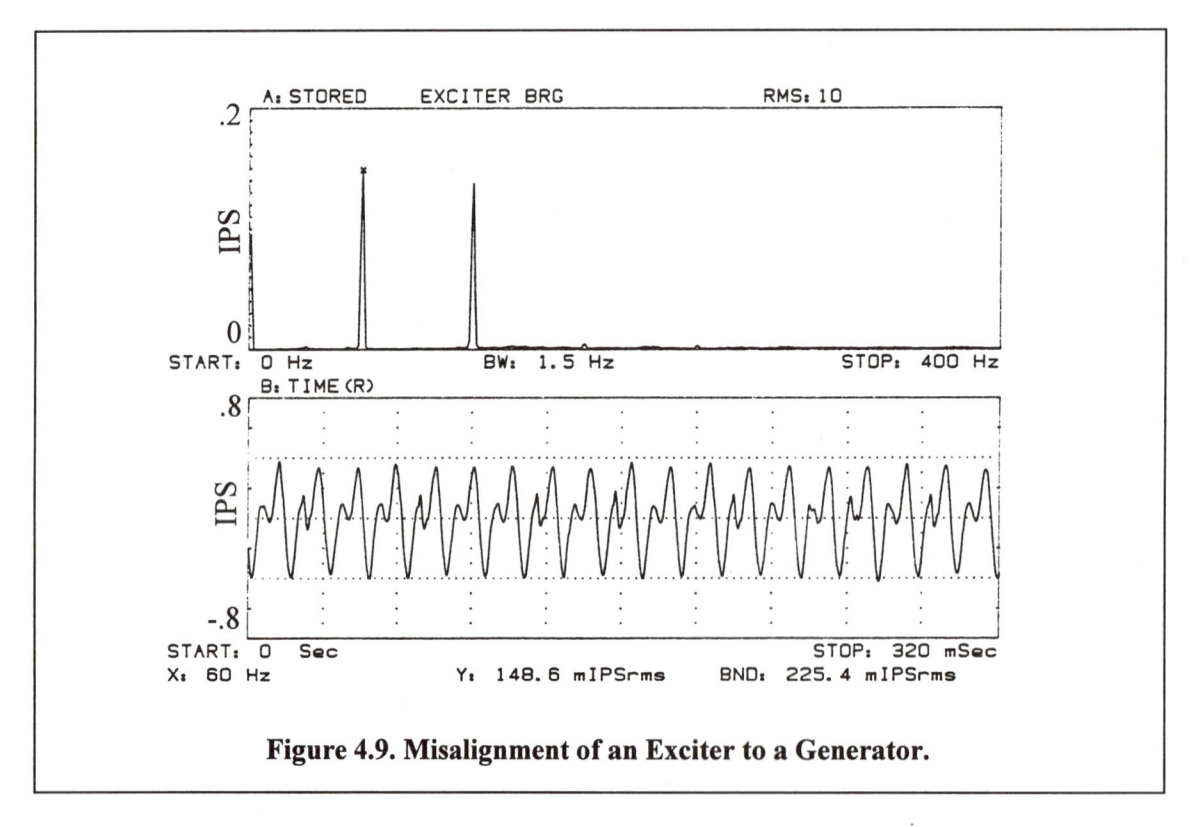

Figure 4.9. Misalignment of an Exciter to a Generator.

misalignment can exceed the first order. The result is an orbit with a figure-eight loop (Figure 4.10). The orbit for minor misalignment is composed largely of once-per-revolution vibration. High first-order axial vibration (out of phase) is also a symptom of misalignment.

Rotor bow or bent shaft. A rotor bow or bent shaft usually causes a preload on the bearings. The center of the mass of a bent shaft can be moved far enough away from the geometric center to cause some mass unbalance. If the machine passes through a critical speed during start-up or coast down, a diagnostic test can be performed to determine the nature of the bow. An abrupt decrease in vibration level (Figure 4.11) followed by an increase is a sign of rotor bow.

Causes of rotor bow include unequal thermal conditions (commonly a result of shorted generator coils), rotor sag, and alteration of metallurgical properties due to rubbing. Compromise balancing can be performed to allow continued operation for a finite period of time. Unequal temperatures at the top and bottom of a rotor that is undergoing convective cooling during the period following shutdown in a steam or gas turbine can distort the rotor and lead to rotor bow. Rotors must be slow-rolled for a period after shutdown to avoid severe vibration when the rotor is restarted. The rotor can be damaged during the process of rolling out the bow.

Rubs can cause bows and unstable conditions below a critical speed because the motion of the shaft is in phase with the forces causing vibration. Heavy rubs and rotor mishandling can cause permanent rotor bows that can sometimes be removed by thermal soaking and peening.

Bearing wear and excessive clearance. Wear or excessive clearance of journal bearings can cause a high once-per-revolution vibration with a stable or unstable phase angle (Figure 4.12). Although the data may seem to indicate mass unbalance, attempts to balance the rotor will fail. Trial weights will cause vibration changes in magnitude and phase angle atypical of a valid balance.

Resonance. Natural frequencies excited by such forces as mass unbalance and its orders amplify vibration. This mechanism is called resonance when it occurs on a structure (see Figure 4.5). The degree of amplification depends on the magnitudes of the force and damping as well as the proximity of the forcing frequency to the natural frequency. Either the forcing frequency (shaft speed) or a natural frequency (dependent upon design) must be changed to solve the problem.

Looseness. Loose mechanical components cause impacts that can be identified in the spectrum as once-per-revolution vibration plus orders (see Table 4.2). The way the machine is supported can cause vibration at 1/4, 1/3, or 1/2 order. Orders close to or at natural frequencies have the highest magnitude because the forces are amplified by resonance. Figure 4.13 shows the vibration of a loose bearing mounting on a fan pedestal.

Eccentricity. Eccentricity of a rotating mechanical component results in vibration at operating speed even though the unit is in balance (Figure 4.14). Pitch-line runout of sheaves (belts) and sprockets (chains) can cause cyclic tightening and loosening of the driver on each revolution of the shaft. Mass unbalance must be eliminated as a cause or the eccentricity must be physically measured (dial indicator) to diagnose the problem correctly.

Distortion. Distortion of a machine casing can cause internal preload of bearings. The result is a vibration at operating speed. Distortion of gearboxes as a result of soft foot causes gear-mesh vibration. Distorted motor casings cause vibration at two times line frequency.

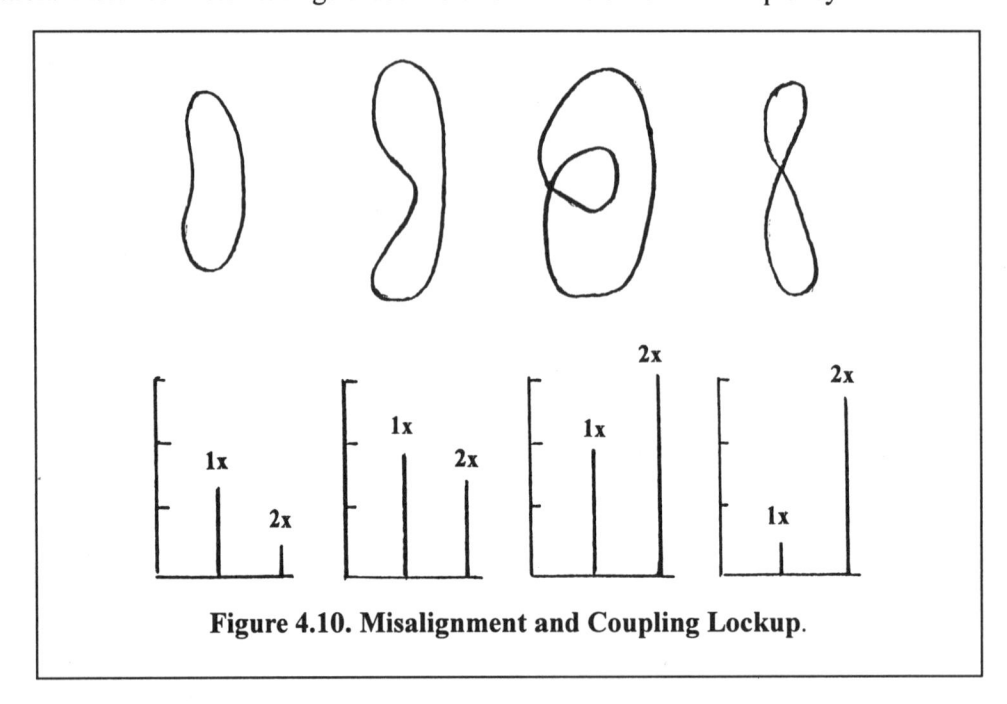

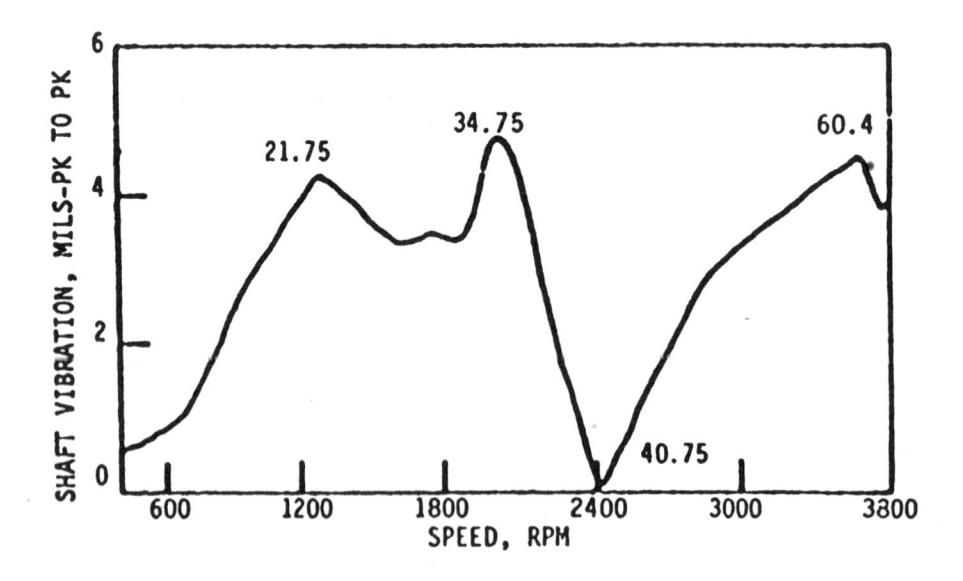

Figure 4.10. Misalignment and Coupling Lockup.

Figure 4.11. Shaft Bow on a 200 Mw Turbine Induced by Coast Down without Oil.

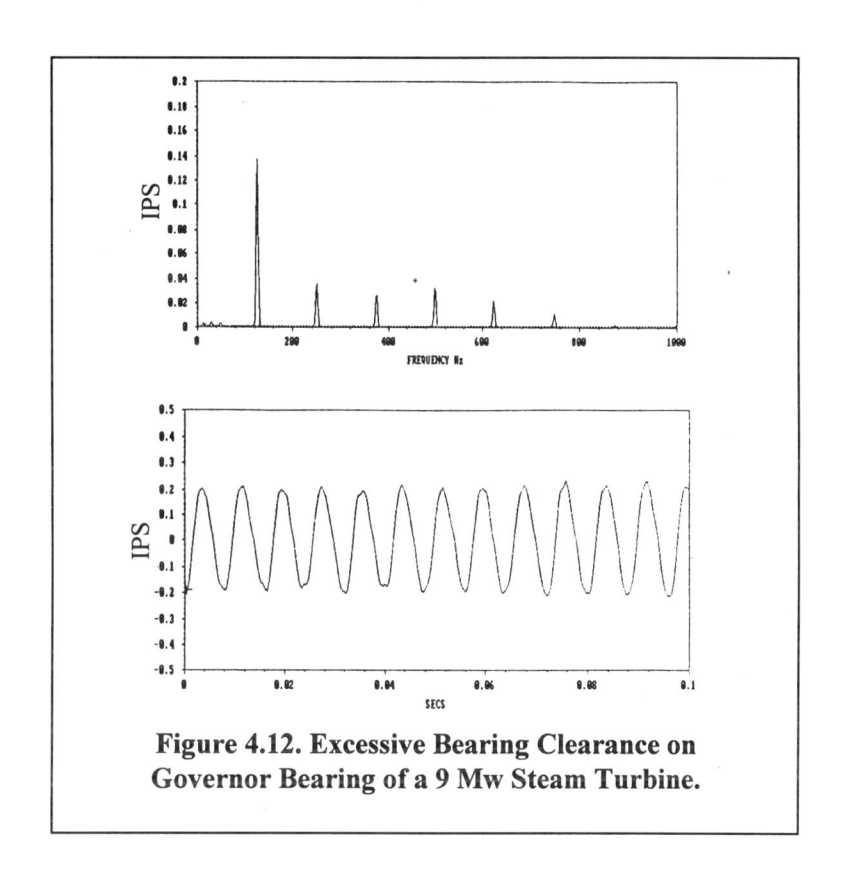

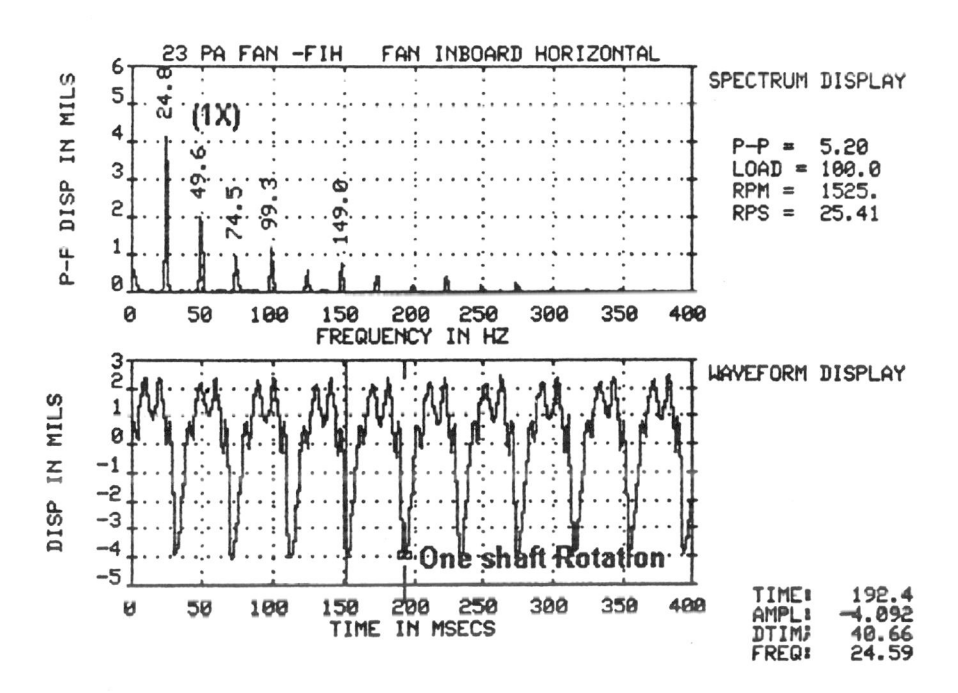

Figure 4.12. Excessive Bearing Clearance on Governor Bearing of a 9 Mw Steam Turbine.

Figure 4.13. Looseness of a Journal Bearing in a Fan.

4.13

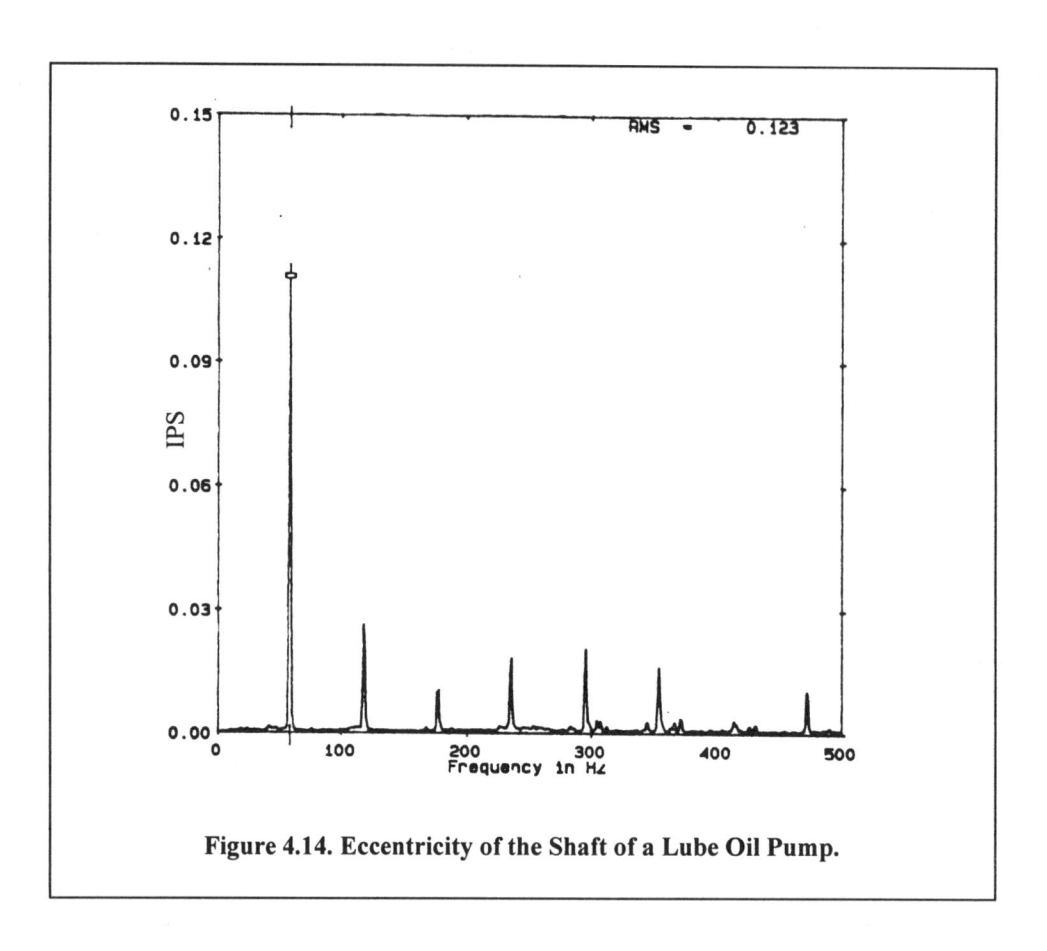

Figure 4.14. Eccentricity of the Shaft of a Lube Oil Pump.

Rolling Element Bearings

When a rolling element passes over a bearing defect in the races or cages (Figure 4.15), pulse-like forces are generated that result in one or a combination of bearing frequencies. The design of the machine determines the magnitude of the vibrations obtained on the bearing housing. Because acceleration levels are low — below 1,000 Hz — acceleration is not a good measure for low-speed machines; that is, machines that operate below 600 RPM. Velocity is the measure of choice in assessing faults below 1,000 Hz.

Bearing frequencies. Rolling element bearings generate frequencies unique to their geometry and operating speed [4.1]. Four basic frequencies can be generated by a defective bearing.

- ball pass frequency of the outer race (BPFO); generated by balls or rollers passing over defective races.

- ball pass frequency of the inner race (BPFI); generated by balls or rollers passing over defective races.

- ball spin frequency (BSF); generated by ball or roller defects.
- fundamental train frequency (FTF); generated by cage defects or improper movements.
- ϕ = contact angle; angle between lines perpendicular to the shaft and from the center of the ball to the point where the arc of the ball and the race make contact (Figure 4.15).
- N = number of rolling elements (balls or rollers).
- P = pitch diameter, in.
- B = ball or roller diameter; average value for tapered bearings, in.
- RPS = speed of rotating unit in revolutions per second.

$$\Omega = RPS = RPM/60$$

$$FTF = (\Omega/2) [1 - (B/P) \cos \phi]$$

$$BPFI = (N/2) \Omega [1 + (B/P) \cos \phi]$$

$$BPFO = (N/2) \Omega [1 - (B/P) \cos \phi]$$

$$BSF = (P/2B) \Omega [1 - (B/P)^2 \cos^2 \phi]$$

The formulas are given in Hz. If cycles per minute (CPM) are desired, use RPM instead of RPS in the formulas. The four bearing frequencies can be modulated by the speed of the rotating unit in RPS to cause sideband frequencies. In somes cases the fundamental train or ball spin frequencies may modulate natural frequencies or ball pass frequencies. The frequencies generated by defective bearings are combinations of bearing frequencies, natural frequencies, and unit rotating frequencies. Pulses can be seen in the time waveform.

Measurement techniques. Vibration analysis of machines for bearing defects encompasses the same principles used for moderate and low-speed equipment. (The one notable exception is aircraft engines.) Measurements can be made with velocity transducers or accelerometers (integrated at low frequencies). All measurements should be made in the load zone as close to the bearing as possible (see Figure 2.12). Radial measurements should be made with radial bearings. Axial measurements may be better for angular contact bearings, depending on the flexibility of the machine. It is necessary to be aware that larger signals from rotating defects such as misalignment, mass unbalance, and gear mesh may overwhelm smaller amplitude bearing frequencies. It is the frequency information that is important. Discrete frequencies occur during the early stage of bearing failure. Sidebands appear during later stages. It is therefore important to use an analyzer with adequate resolution to determine the operating frequency and the sidebands of the fundamental train frequency. The time waveform contains pulses in the early stages of bearing failure.

> **Example 4.1**: Frequency calculation for rolling element bearings.
>
> Bearing No. SKF 230-600
>
> B = 2.598 in.
> P = 29.11 in.
> ϕ = 8.166°
> N = 29
> Ω = RPM/60 = 213/60 = 3.55 RPS
> (B/P) cos ϕ = (2.598/29.11) 0.9899 = 0.0883
>
> BPFO = (29/2) 3.55 [1 - 0.0883] = 46.9 Hz
>
> BPFI = (29/2) 3.55 [1 + 0.0883] = 56 Hz
>
> FTF = (3.55/2) [1 - 0.0883] = 1.6 Hz
>
> BSF = [29.11/(2) (2.598)] 3.55 [1 - 0.0883^2] = 19.7 Hz

Analysis techniques. The narrow-band vibration analysis of rolling element bearings involves several steps.

- Calculating bearing frequencies.
- Measuring and analyzing vibration signals.
- Identifying sidebands and center frequencies in the spectrum.
- Evaluating the spectrum and time waveform for shape, energy, and amplitude.

Defects. Table 4.3 is a synopsis of bearing defects and their symptoms. Figure 4.3 and Figures 4.16 through 4.19 are examples of the symptoms of bearing defects described in the Table. A small defect on the inner or outer race of a bearing produces discrete spectral lines at the appropriate bearing frequency and its orders. The BPFI of 42 Hz (see Figure 4.16) and its harmonics of 84 Hz, 124 Hz, and 166 Hz were measured on a bearing that showed shallow flaking. The vibration from the roller passing through the defect can be seen in the time waveform. Sidebands appear as the condition of the bearing deteriorates. Figure 4.17 is a spectrum of a bearing that failed two weeks after it was analyzed. Note the center bearing frequencies and their orders surrounded by sidebands. The sidebands are the speed of the shaft. Figure 4.18 is a spectrum of a bearing with a cage defect. The FTF of 6 Hz and its harmonics modulate natural frequencies in the unit at 78 Hz, 151 Hz, and 224 Hz. Even though the amplitude is low at 0.03 IPS_{rms}, the condition of the bearing is critical, and it should be removed.

Table 4.3. Analysis of Rolling Element Bearing Defects.

Defect or Condition	Frequency	Time Waveform/ Spectrum Shape	Comment	Figure #
outer race defect	BPFO and multiples	multiples of BPFO	shallow flaking increasing in severity after one year	4.3
inner race defect	BPFI and multiples	decreasing size harmonics	shallow flaking	4.16
inner race defect	BPFI and multiples	decreasing size haronics modulated by operating speed	bearing lasted 14 days	4.17
ball defect	BSF or FTF and multiples	natural frequencies modulated by FTF	balls rattle bearing at natural frequency	4.18
excessive internal clearance	natural frequencies	multiples of RPS modulate natural frequencies	bearing showed no defects/showed excessive wear	4.19

Excessive bearing clearance can appear as multiples of operating speed, or, if the clearance is sufficiently large, natural frequencies of the unit are excited (Figure 4.19). The bearing of Figure 4.19 had no visible defects but emitted a loud noise during operation.

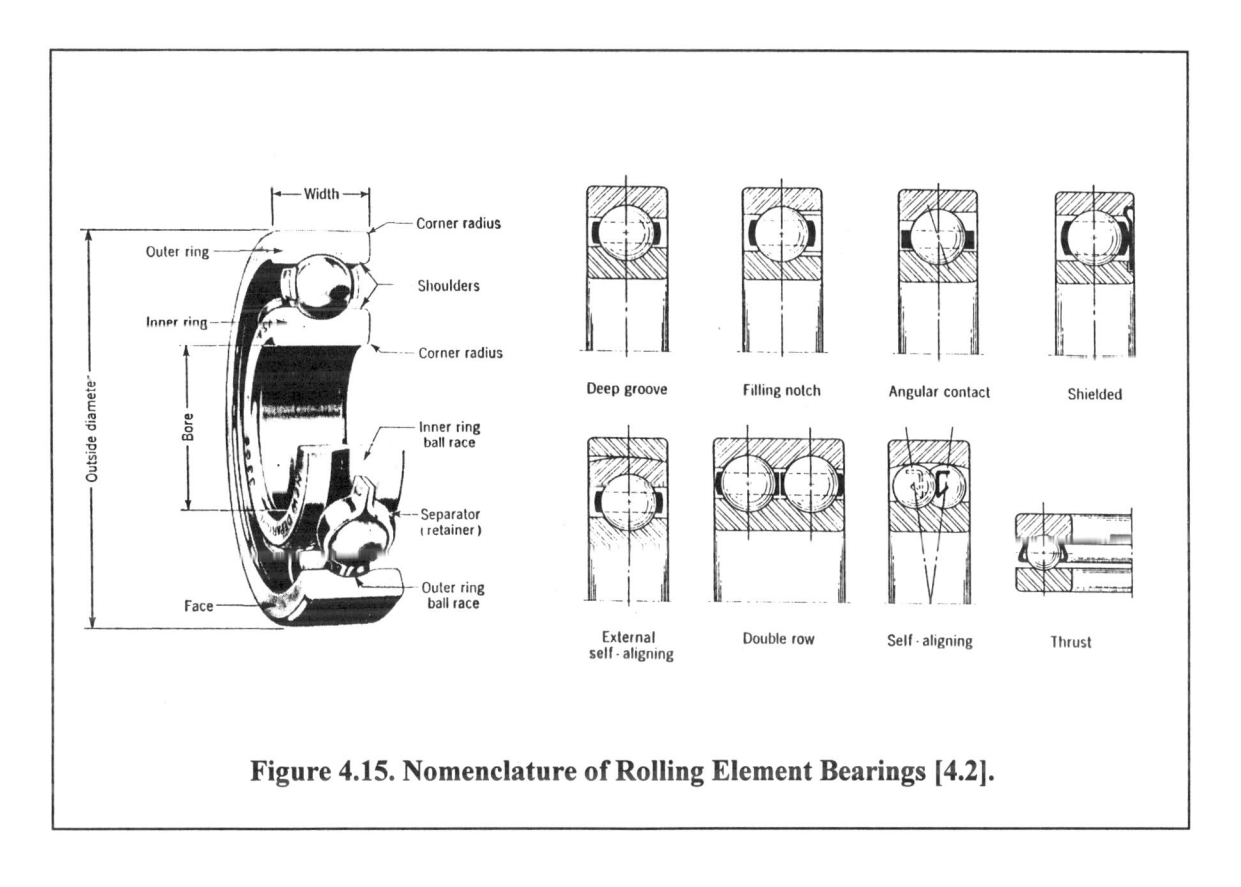

Figure 4.15. Nomenclature of Rolling Element Bearings [4.2].

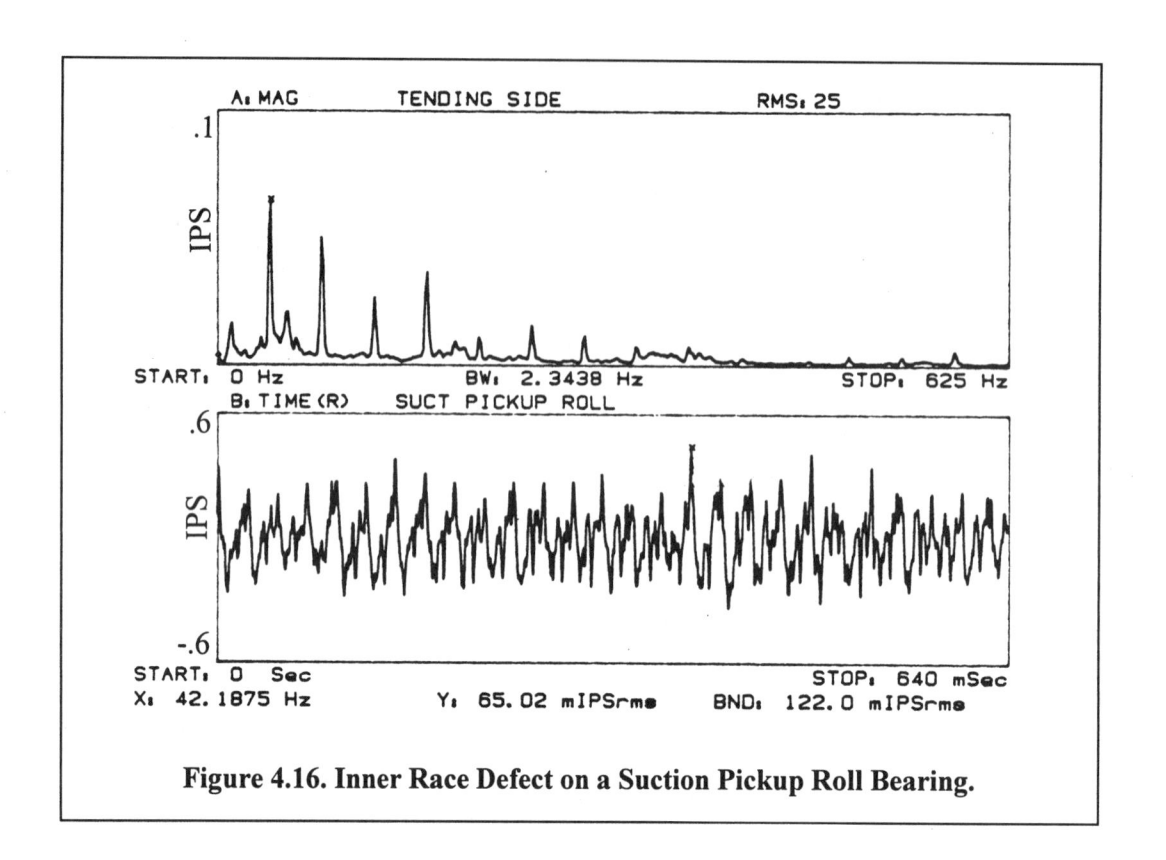

Figure 4.16. Inner Race Defect on a Suction Pickup Roll Bearing.

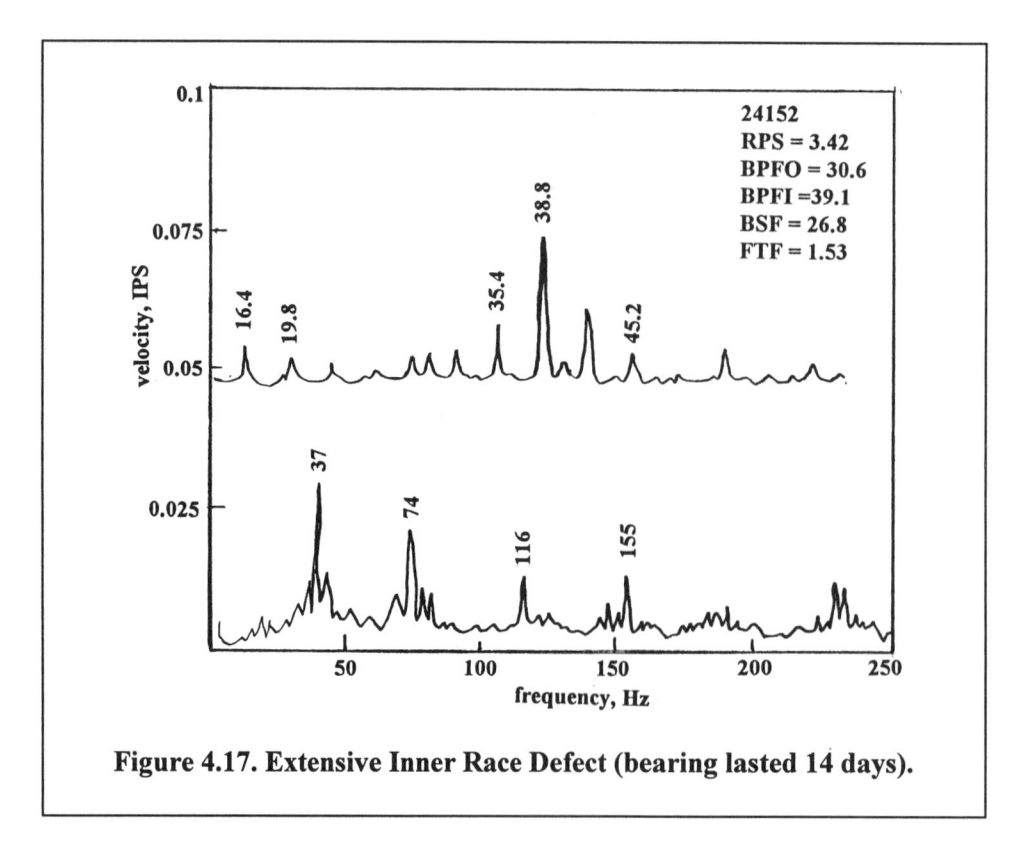

Figure 4.17. Extensive Inner Race Defect (bearing lasted 14 days).

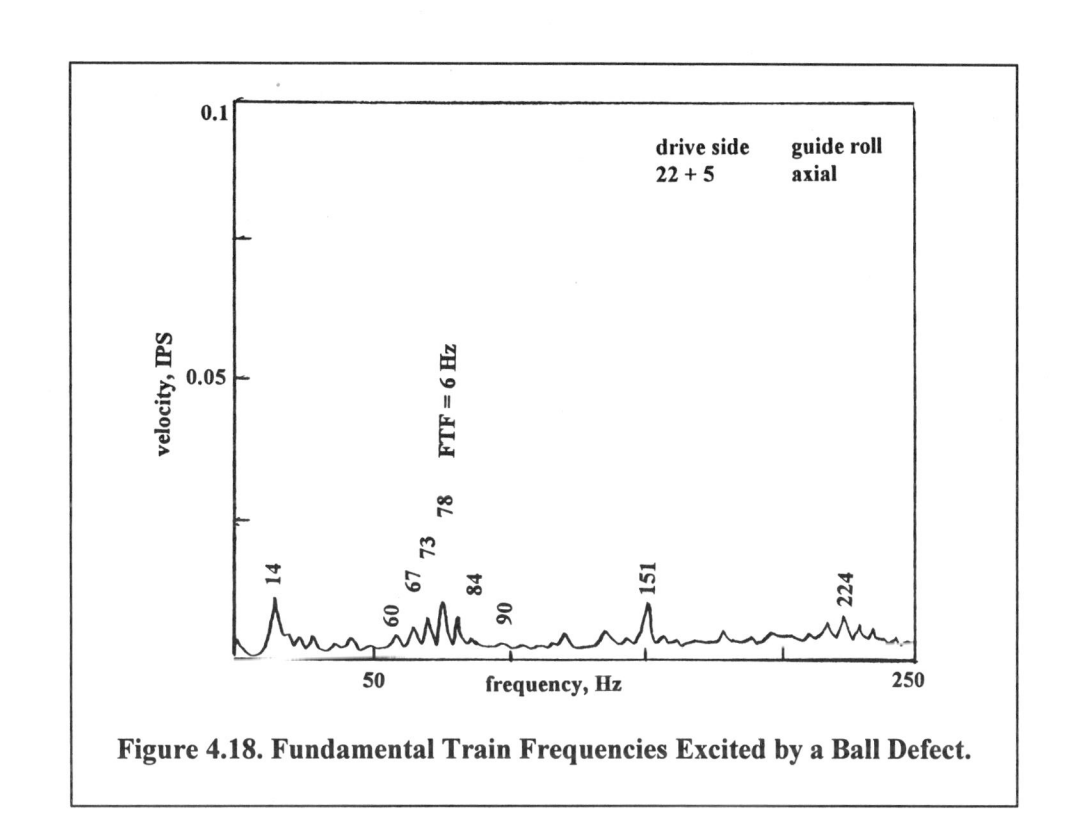

Figure 4.18. Fundamental Train Frequencies Excited by a Ball Defect.

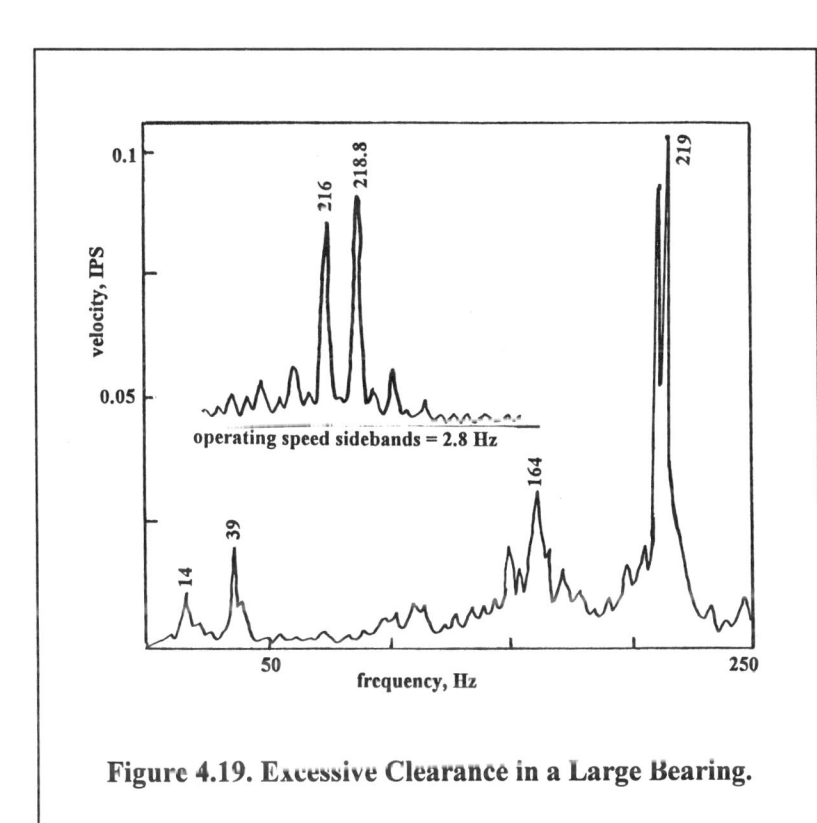

Figure 4.19. Excessive Clearance in a Large Bearing.

High-frequency detection (HFD) methods. Acceleration is the primary measure for condition evaluation and diagnosis with PC based data collection systems. High-frequency signal-processing techniques are utilized in data collectors and digital signal processing systems. These techniques are of value in detecting the earliest stages of bearing failure because of the physics of rolling element bearing degradation. The techniques are also applied to condition evaluation of low-speed equipment, in which structure and mass are such that overall surveillance does not adequately reflect changes in machine condition.

High-frequency detection methods include two types of measurements. One is the single value summation of energy over a filtered range of vibration, typically 5 kHz to 60 kHz. The second is a spectral representation of a filtered signal that has been amplified and demodulated with respect to amplitude and/or frequency. The purpose is to visualize repetitive information so that it can be evaluated as a traditional spectrum, either HFD or an envelope.

High-frequency response must be measured with an accelerometer. The upper limits of the linear measurement range are from 5 kHz to 10 kHz and as high as 25 kHz, depending on the accelerometer used. Because the natural frequency of the accelerometer is always higher than the linear measurement range, some signal amplification is provided through the range of the natural frequency response when overall measurements are used.

Single value readings are the most common ones used in conjunction with a computer-based predictive maintenance system and are most effective when trended at specific points on the machine. The range of values changes with the accelerometers, even with the same product model. Data are more consistent if the transducer mounting is consistent. A magnetic mount provides the most cost-effective data. Single-value data must be reconciled with machine speed — i.e., higher speeds generate higher values. No direct diagnostic information is available from single-value readings. The analyst must base his judgments on the reading, the history of the machine, and other measurements, whether or not they are vibration-based.

The following mechanisms and conditions can cause high readings. This list is not all inclusive. Included are impacts, rubs, inadequate lubrication, flow turbulence in pump systems, poor mechanical seal conditions, high-pressure leaks (steam/air), and preloads/improper interference fits. The most cost-effective action when high-frequency measurement is elevated is bearing lubrication.

Gearboxes

Gearboxes generate high-frequency vibrations as a result of the gear-meshing function of the box. The greater the number of gear teeth in mesh at any instant the smoother is the performance of the box. Gearbox faults and their sysmptoms are summarized in Table 4.4. Both the time waveform and the spectrum must be analyzed. Deterioration of condition complicates fault diagnosis.

The fact that pulses observed in the time waveform identify as broken gear teeth was first observed by Taylor [4.1]. Other impacts such as binding of gear teeth can also cause pulses in the time waveform. Gear-mesh frequencies with sidebands at operating speeds identify such problems as gear-mesh wear and gearbox distortion.

Before any fault is analyzed, gearbox frequencies must be calculated from vendor-supplied data. Figure 4.20 shows a double-reduction gearbox. The frequencies involved in the gearbox are calculated in Example 4.2.

Gear-mesh problems attributable to uneven wear, improper backlash, scoring, and eccentricity generally appear in the spectrum as gear mesh with sidebands at the frequency of the speed of the faulty shaft. Badly worn gears will show multiples of gear-mesh frequency with sidebands. The best signal strength for herringbone and helical gears is usually obtained from an

Table 4.4. Identification of Malfunctions in Gears and Gearboxes.

Fault	Frequency	Example (Figure #)	Spectrum Time Waveform
eccentric gears	gear mesh	4.21	gear mesh with sidebands at frequency of eccentric gear
gear-mesh wear	gear mesh	4.2	gear mesh with sidebands at frequency of worn, scored, or pitted gear(s); sometimes 1/2, 1/3, 1/4 harmonics of gear mesh
improper backlash of end float	gear mesh	4.6	gear mesh with orders and sidebands at frequency of pinion or gear
broken, cracked, or chipped gear teeth	natural frequencies	4.22	pulses in time waveform; natural frequencies in spectrum
gearbox distortion	gear mesh and/or natural frequencies	4.23	gear mesh and orders in spectrum; varying gear-mesh amplitude in time waveform – shaft frequency plus low-amplitude orders

Figure 4.20. Schematic Diagram of a Double-Reduction Gearbox.

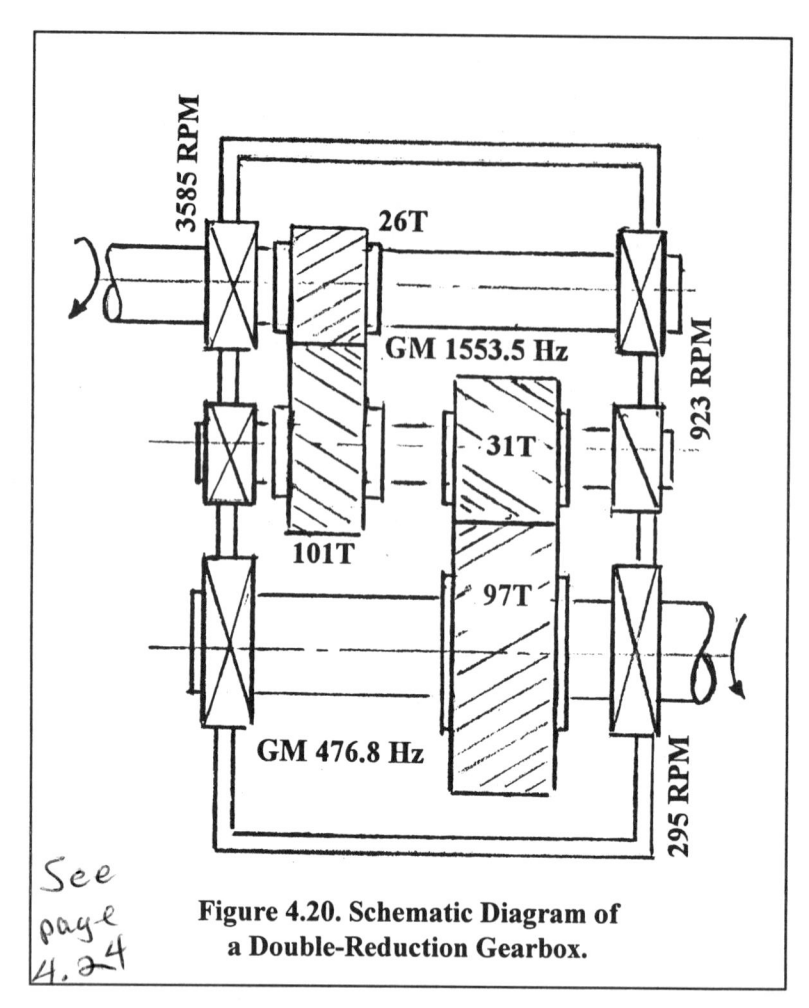

(Diagram labels: 3585 RPM, 26T, GM 1553.5 Hz, 923 RPM, 31T, 101T, 97T, GM 476.8 Hz, 295 RPM)

See page 4.24

axial measurement. Gearboxes with spur gears should be measured in the radial direction. The gearbox shown in Figure 4.21 has an eccentric pinion. Sidebands at pinion speed (1,800 RPM) can be observed on the gear-meshing frequency (730 Hz). The data shown in Figure 4.2 were taken from a large single-reduction gearbox with worn gears. Note the 1/2 order gear mesh. This gearbox has a common factor of two between the teeth. The data shown in Figure 4.6 were taken from a double-reduction, right-angle gearbox with an input speed of 1,776 RPM and a bevel gear set (gear mesh 730 Hz) and a helical low-speed gear set (gear mesh 466.6 Hz). The end float/backlash problem of this unit caused multiples of high-speed gear mesh with sidebands at the input shaft speed. The time waveform provides the best information for identifying broken, cracked, or chipped teeth [4.1]. Pulses appear at a frequency equal to the number of defective teeth multiplied by shaft speed (Figure 4.22) unless more than one faulty tooth is in mesh simultaneously. In this instance a tooth is chipped on the pinion, resulting in one pulse every 46.5 milliseconds. Problems with misalignment and distortion are generally identified in the time waveform as modulation of the gear-mesh frequency (Figure 4.23). The binding and releasing of the gear mesh shown in the time waveform are identified as difference frequencies in the spectrum at the speed of the input shaft.

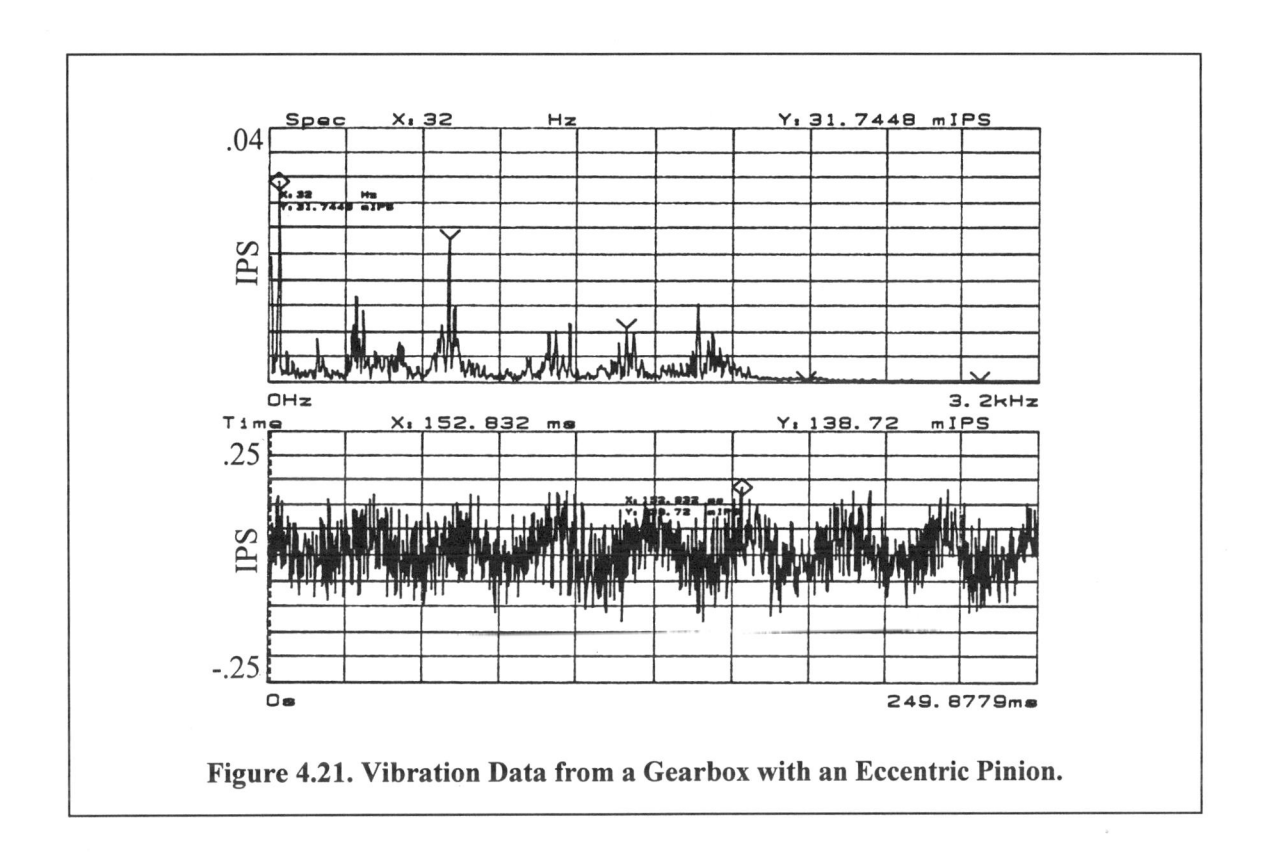

Figure 4.21. Vibration Data from a Gearbox with an Eccentric Pinion.

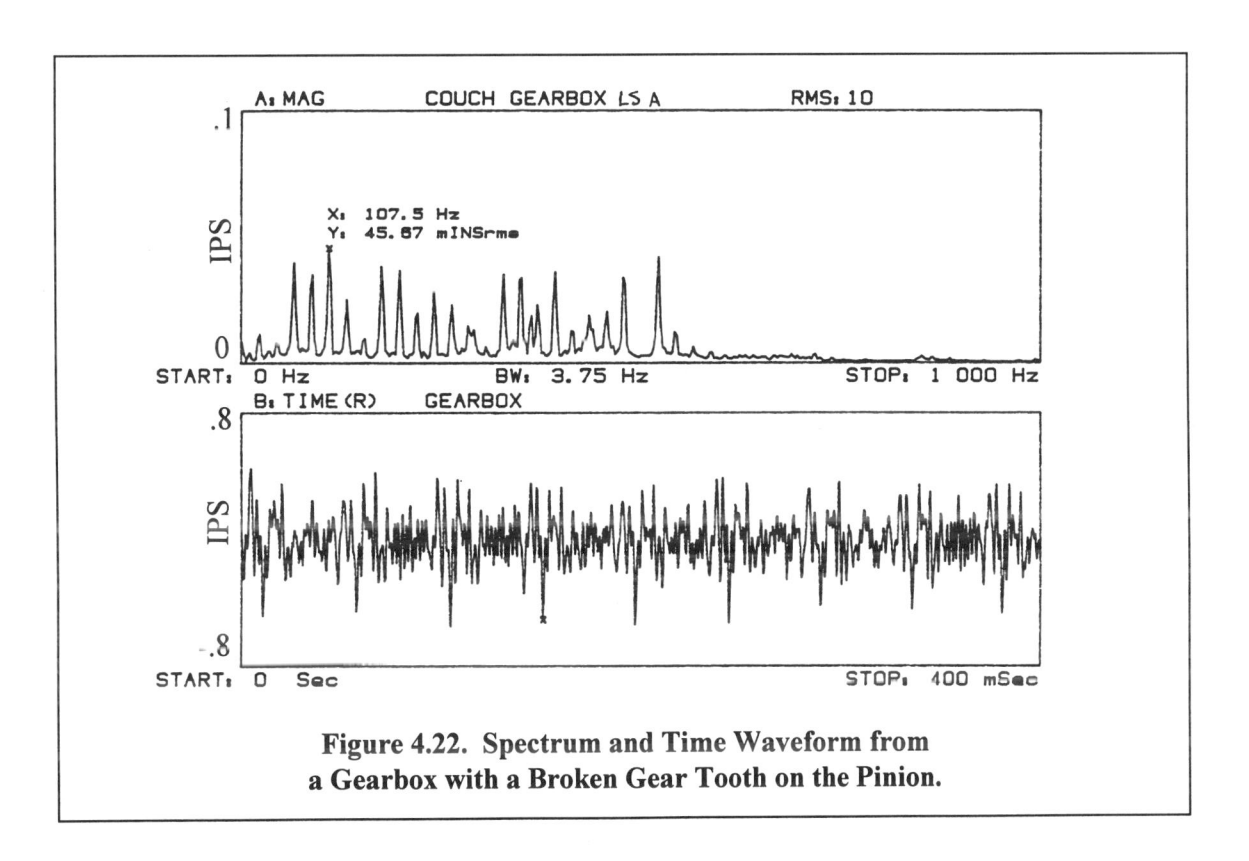

Figure 4.22. Spectrum and Time Waveform from a Gearbox with a Broken Gear Tooth on the Pinion.

Figure 4.23. Data from a Misaligned and Distorted Gearbox.

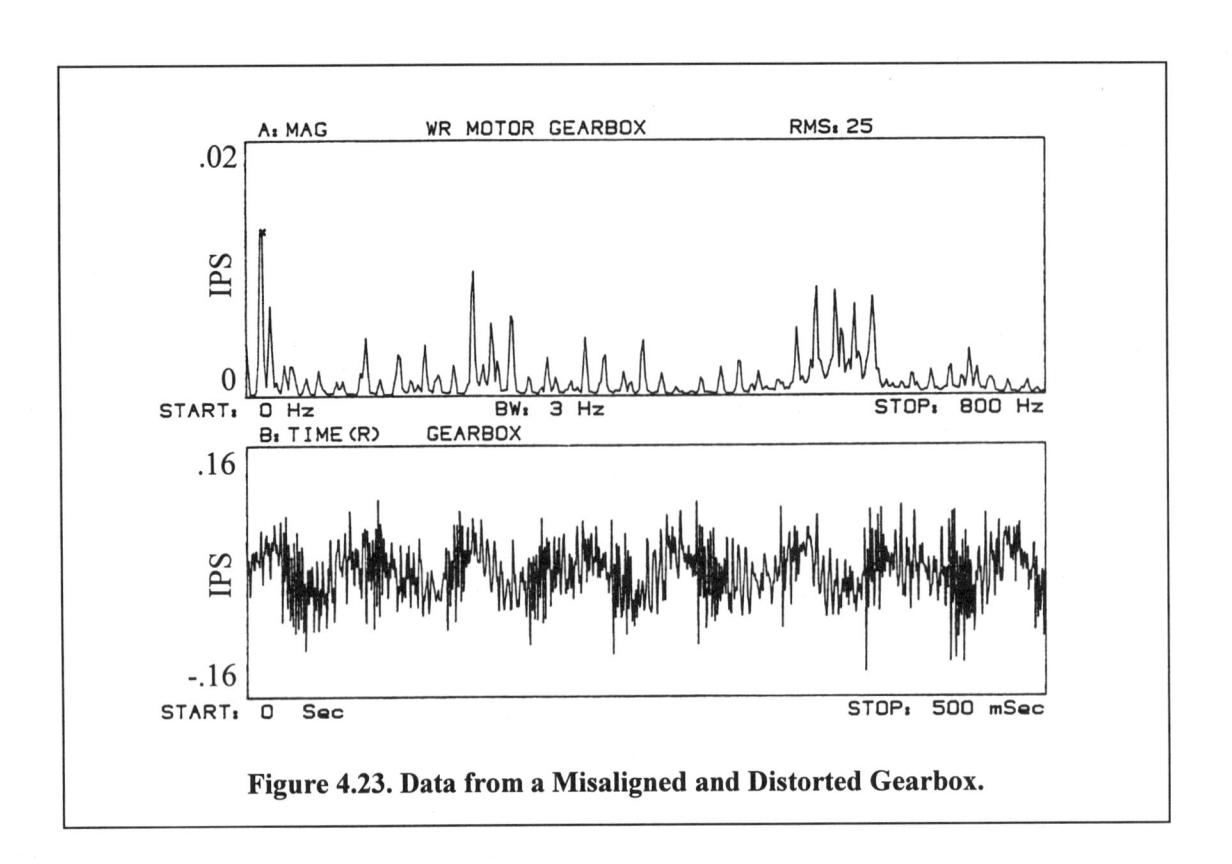

> **Example 4.2**: Gear frequency calculations.
>
> The gearbox in Figure 4.20 is driven by a two-pole motor at 3,585 RPM. What are the gear-mesh frequencies and shaft speeds?
>
>   input shaft speed = 3,585 RPM
>
>   intermediate shaft speed = (3,585 RPM) [(26 T)/101 T] = 923 RPM
>
>   output shaft speed = (923 RPM) [31 T/97 T] = 295 RPM
>
>   high-speed gear mesh = (3,585 RPM) (26 T) = 93,210 CPM, or 1,553.5 Hz
>
>   low-speed gear mesh = (922.87 RPM) (31 T) = 28,609 CPM, or 476.8. Hz

Electric Motors

The induction motor is driven by a voltage at line frequency of 60 Hz (in the U.S.A.) directly from the power terminal or by a controller that reforms the power to a different line frequency that provides variable speed. Induction motors are designed to operate at a number of fixed speeds by the number of poles. The relationship between synchronous motor speed (no load), number of poles, and line frequency is expressed in the following simple equation. The synchronous motor speed is the frequency of the magnetic field.

synchronous motor speed (SMS) = 2 times line frequency/number of poles

Example 4.3: Synchronous motor speed calculation.

What is the synchronous speed of an 8-pole induction motor operating on 60 Hz power?

SMS = (2) (60)/8 =
(15 cycles/sec) (60 sec/min) = 900 RPM

Example 4.4: Calculation of slip frequency for an induction motor.

A 4-pole induction motor operates at 1,774, RPM. What is the slip frequency in Hz?

SMS = (2) (60)/4 = 30 Hz

slip frequency = 30 Hz - (1,774/60)
= 0.433 Hz, or 26 CPM

An induction motor slips — that is, it does not run at synchronous motor speed — because of its load. The difference between synchronous motor speed and the actual motor speed is called slip (see Example 4.4).

Figure 4.24 is a cut-away section of an induction motor. The mechanical malfunctions that affect rotating machines also cause problems in electric motors. Included are mass unbalance, looseness, resonance, misalignment, eccentricity, bearing defects, and distortion. In addition, electric motors are sensitive to common mechanically-induced electrical faults that generate mechanical vibrations (Table 4.5). Included are air-gap variation, failure to stay on magnetic center, stator flexibility, broken or loose rotor bars, and shorted laminations. A complete chart on causes, checks, and cures for mechanical and electrical problems in alternating current electrical motors has been published [4.3]. Electrical malfunctions in the stator induce vibrations at two times line frequency (120 Hz) and sidebands at the number of poles times slip frequency. Vibrations at frequencies equal to the number of slots of rotor or stator times motor speed occur with an eccentric rotor or if the number of slots is similar in stator and rotor. Broken rotor bars create operating-speed vibration with sidebands at the number of poles times slip frequency.

The most typical abnormal vibration in two-pole induction motors is associated with air-gap variation. A motor with an unbalanced or eccentric armature or some other mechanical condition, (e.g., flexible stator) that causes an air-gap variation with rotation generates vibration at exactly two times line frequency (Figure 4.25). If a mechanical component of vibration is present

4.25

at two times operating speed, beats occur because the frequencies of the two components are very close. Electrical problems obviously disappear when the machine is shut down. It is best to shut down the motor and immediately observe the vibration spectrum.

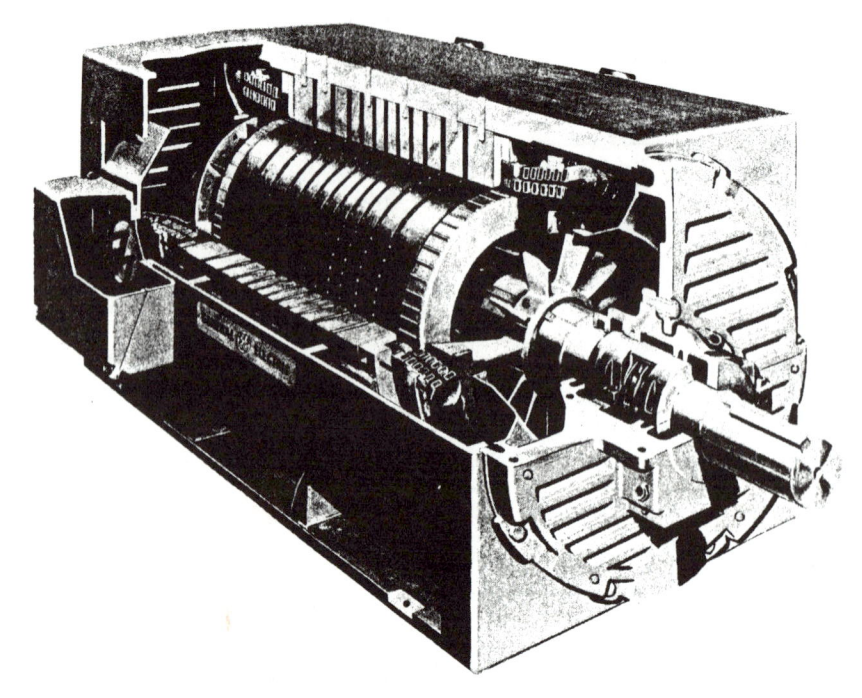

Figure 4.24. Cut-Away Diagram of an Induction Motor.
Courtesy of General Electric Company

Such faults as a broken or loose rotor bar or a shorting ring connection to the rotor bar are evident only when the motor is under load. Broken rotor bars cause sidebands equal to the number of poles multiplied by the slip frequency on the operating speed. Note the sidebands on the operating speed of the data shown in Figure 4.26. Shorted laminations cause local hot spots in the rotor that in turn cause the rotor to bow. The result is vibration at operating speed.

Figure 4.27 contains radial data from a motor with a rotor that is eccentric at the bearings. Both sidebands on the operating speed and the 2x line frequency can be seen in the spectrum. A motor with high axial vibrations at nonsynchronous frequencies (Figure 4.28) is being held off magnetic center. As the rotor works to get on magnetic center (axially) impacting occurs that generates the natural frequencies. Radial vibration at two times line frequency and multiples is a sign of shorted windings in the stator (Figure 4.29). The 12-pole induction motor has an abnormally high vibration level as a result of stator shorts.

Table 4.5. Identification and Correction of Motor Malfunctions — Electrical Effects.

Fault	Frequency	Example (Figure #)	Spectrum; Time Waveform/Orbit Shape	Correction/Comment
air-gap variation *eccentricity*	120 Hz	4.25	120 Hz plus sidebands, beating 2x with 120 Hz	center armature relieving distortion on frame; eliminate excessive bearing clearance and/or any other condition that causes rotor to be off center with stator
broken rotor bars −1772.24 +1775.76	1x RS 1774 RPM .44 × 4 = 1.76	4.26	1x and sidebands equal to (number of poles x slip frequency) or the fpp	replace loose or broken rotor bars
eccentric rotor	1x	4.27	1x, 2x/120-Hz beats possible	may cause air-gap variation
stator flexibility	120 Hz		2x/120-Hz beats	stiffen stator structure
off magnetic center	1x, 2x, 3x	4.28	impacting in axial direction	remove source of axial constraint-bearing thrust, coupling
stator shorts	120 Hz and harmonics	4.29	120 Hz and harmonics	replace stator

Figure 4.25. Vibration Data from a 4,000 HP Electric Motor with an Air-Gap Problem.

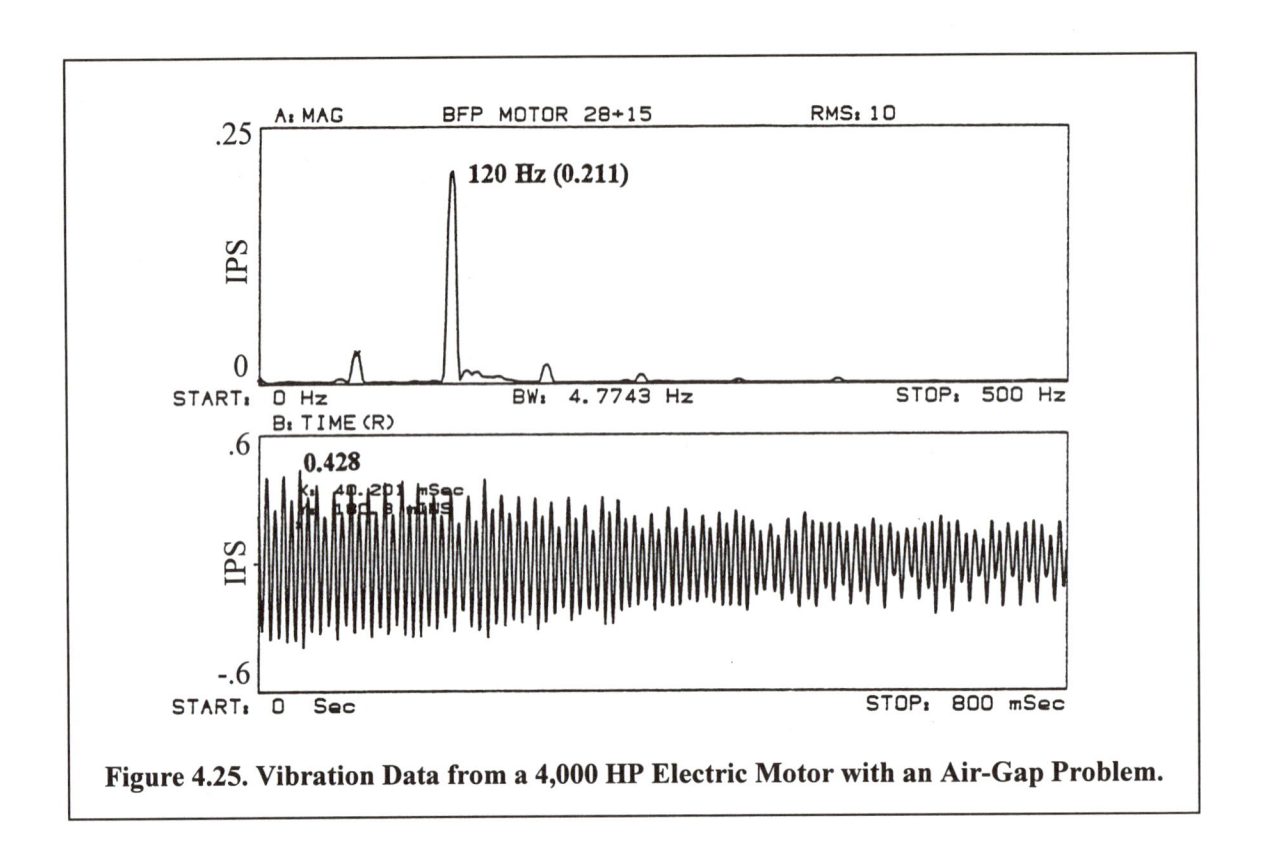

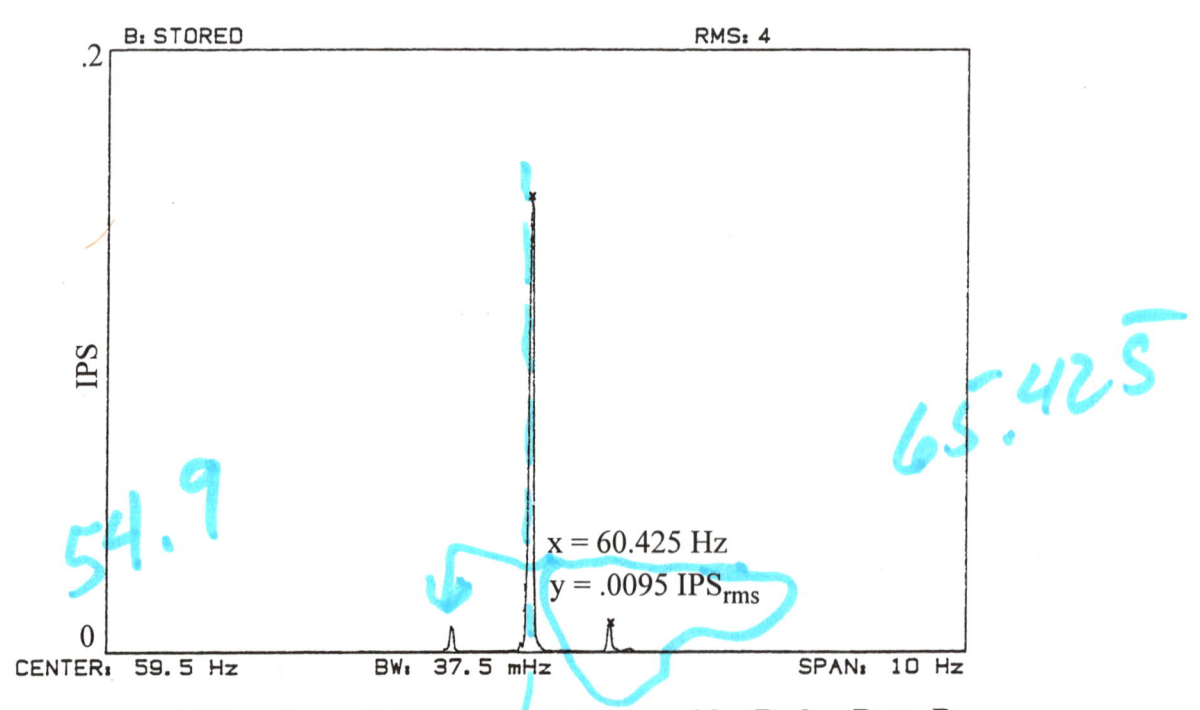

x = 60.425 Hz
y = .0095 IPS$_{rms}$

Figure 4.26. Data from a 2,000 HP Electric Motor with a Broken Rotor Bar.

Figure 4.27. Data from a 1,000 HP Induction Motor with an Eccentric Rotor.

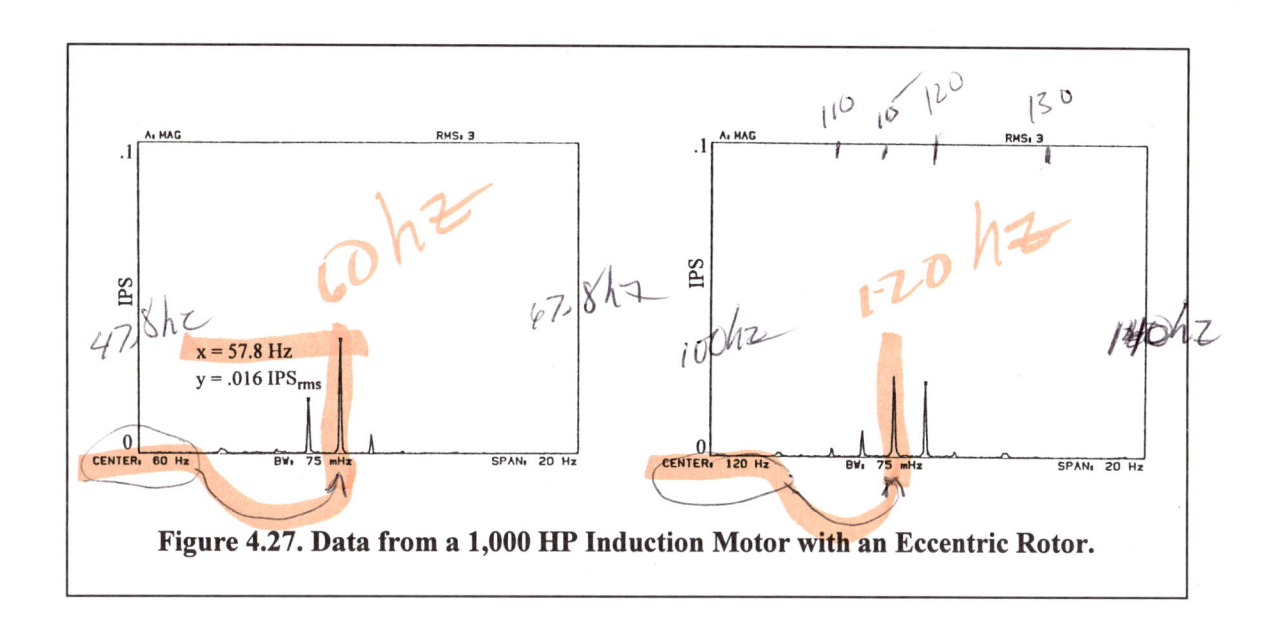

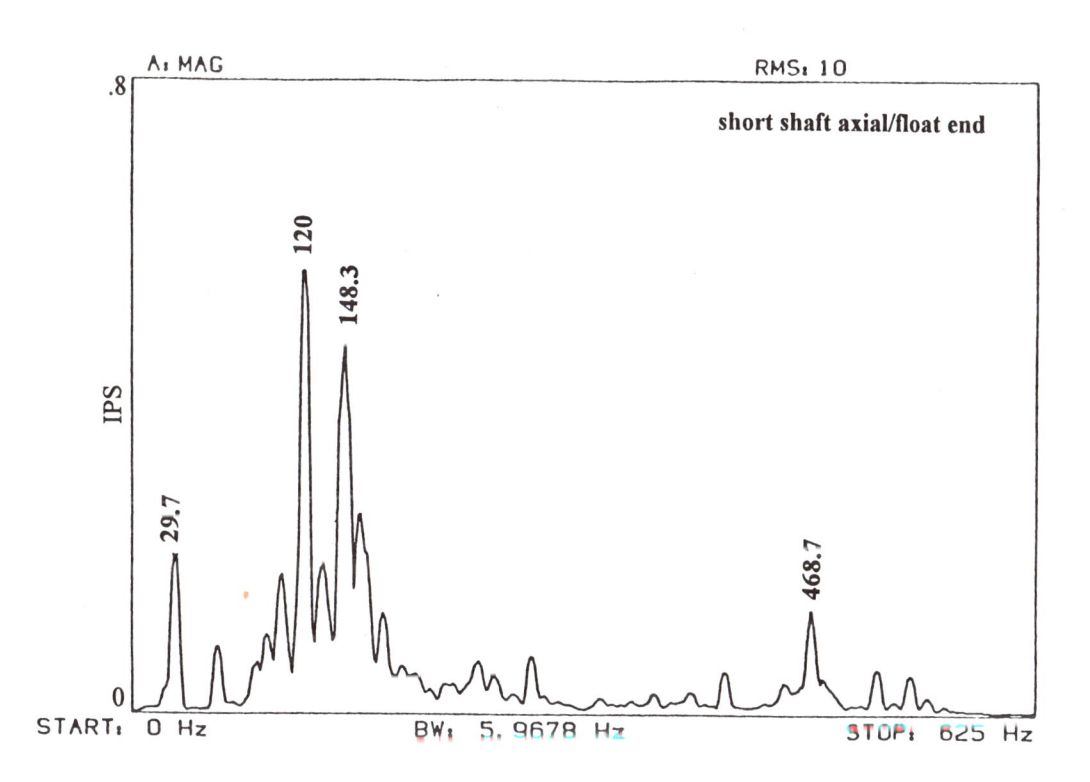

Figure 4.28. Axial Vibration from a 1,200 HP Induction Motor with Rotor Held Off Magnetic Center.

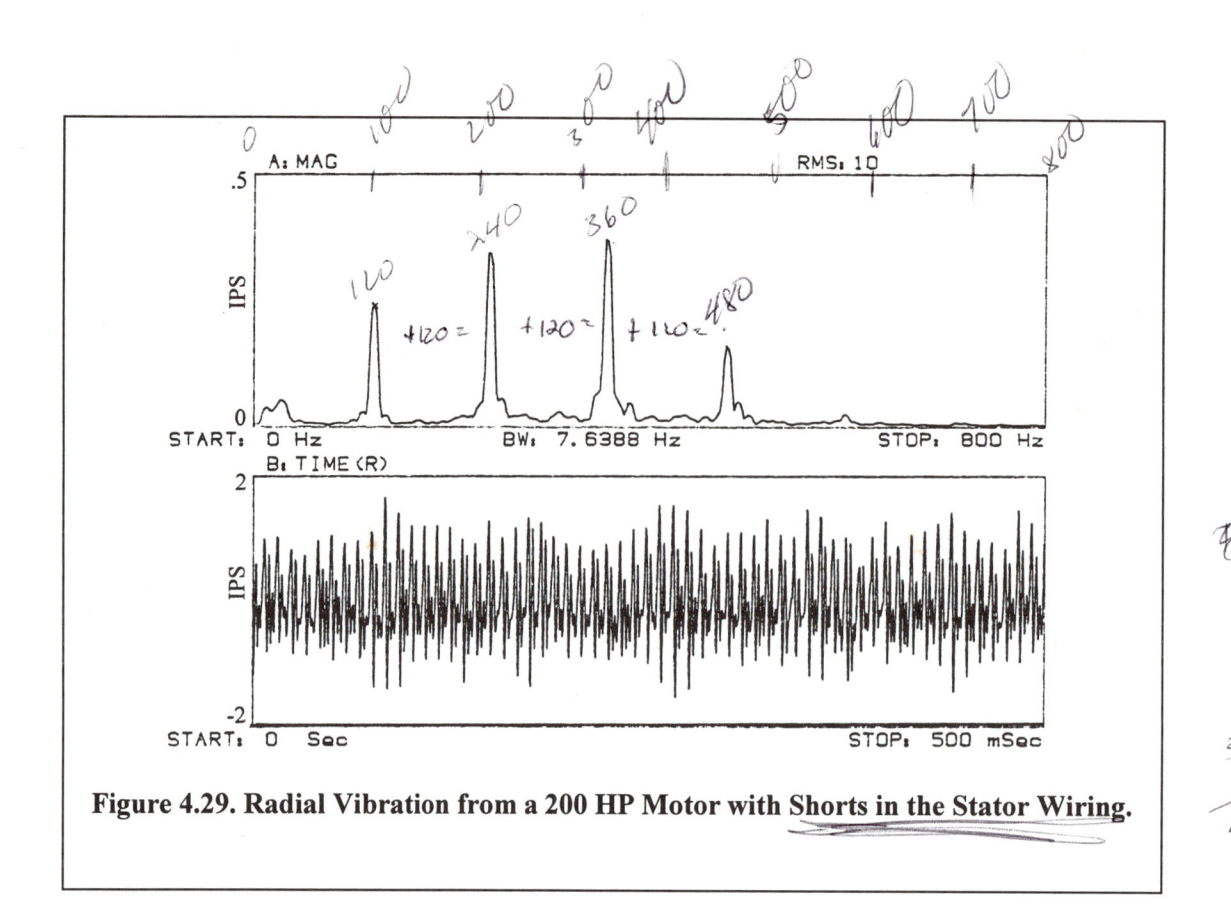

Figure 4.29. Radial Vibration from a 200 HP Motor with Shorts in the Stator Wiring.

Centrifugal and Axial Machines

Pumps and fans transport fluids by converting mechanical work into energy of the fluid in the form of pressure and velocity. Compressors increase the energy of the compressed fluid as pressure. These machines are driven by either electric motors or turbines (gas and steam). Pumps, fans, and compressors are either radial flow (centrifugal) or axial flow, depending on the motion of the flow as it passes through the impeller. In a pump the working fluid is a liquid. The working fluid is a gas in fans and compressors. Fans are distinguished from compressors by the density change (compression) in the moving fluid induced by the the compressor. The fluid moved by a fan experiences very little compression. The performance characteristics of all centrifugal and axial machines are related to the head (pressure) and the efficiency and horsepower of the flow rate of the fluid. Figure 4.30 shows characteristic curves for centrifugal machines with different blade curvatures. Stable and efficient operation requires that the machine operate on the negative slope of the curve. Otherwise, unstable flow causes excessive vibration that is hydraulically or aerodynamically induced.

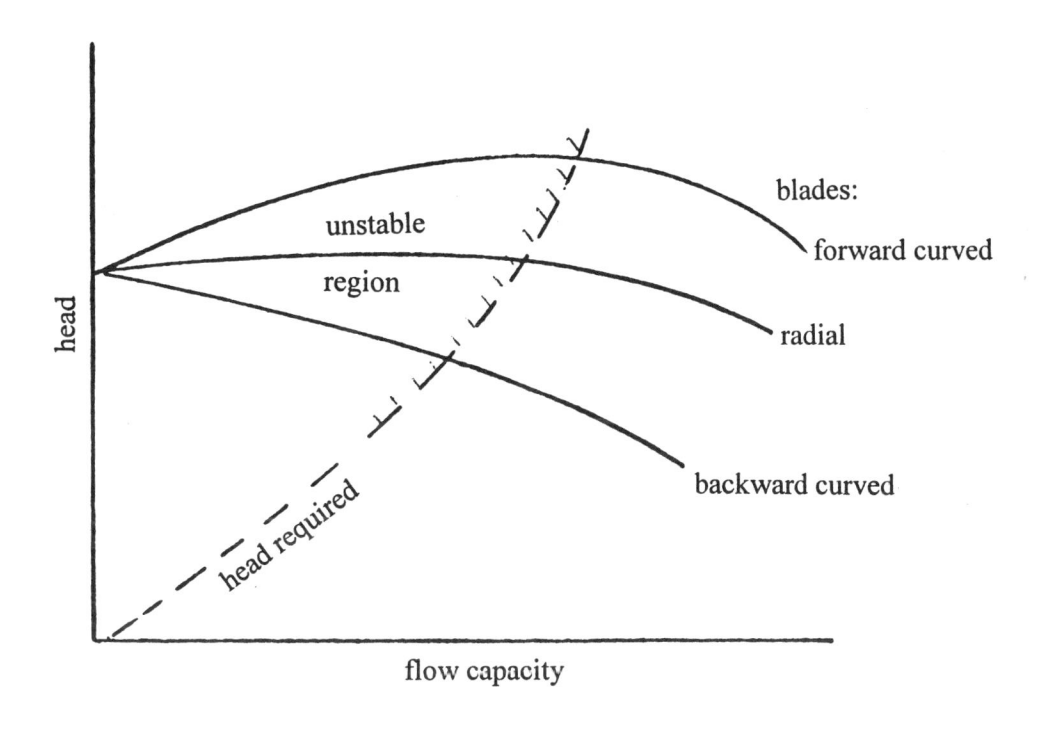

Figure 4.30. Characteristic Curves for Centrifugal Machines.

Pumps

A centrifugal pump consists of rotating elements (shaft and impeller) and stationary elements (casing, bearings, and stuffing boxes). Wear rings are used on multistage pumps to increase efficiency. The liquid to be pumped is forced into a set of rotating vanes by atmospheric or other pressure. The rotating vanes discharge the fluid to the periphery of the pump at a higher pressure and velocity. Most of the velocity is converted to pressure in the casing volutes or diffusers. Impellers are classified as single or double (axial balance) suction. Table 4.6 is a list of pump faults.

Table 4.6. Common Pump Faults.

Critical Speeds
Structural Resonances (principally vertical pumps)
Acoustical Resonances (piping design)
Impeller Eccentricity (nonconcentric machining, deflection of impeller shaft due to head)
Impeller Balance
Impeller/Diffuser Clearance (gaps)
Recirculation (low flow)
Cavitation (low suction head)
Oil Whirl (bearing design and excessive clearance)
Wear Ring Clearance (modifies critical speeds, may induce oil whirl)

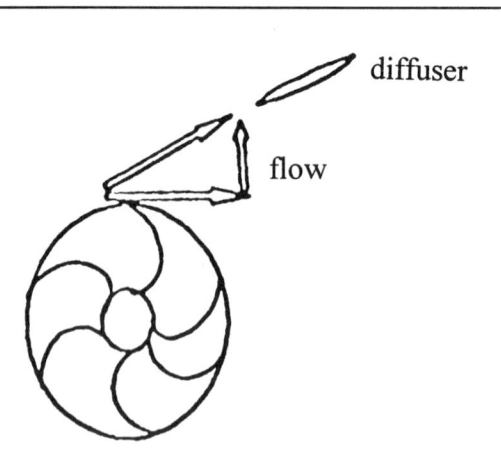

At best efficiency design point, fluid discharge angle matches angle of diffuser and flow is smooth with minimal disturbance.

If flow is decreased (too much back pressure) or is increased (too little back pressure), the fluid angle no longer matches the fluid angle, resulting in higher vibration and a loss of efficiency.

Figure 4.31. Flow Path Characteristics.

The fluids that pumps transport are not compressible. It is thus possible that large interactive forces will be transmitted between the rotating and the stationary components (Figure 4.31). In addition, under certain conditions the liquid can vaporize and then collapse back into a liquid state, causing shock waves that can destroy the impeller of the pump. This process is known as cavitation. The presence of abnormal interactive forces and cavitation are functions of pump operation relative to design conditions (Figure 4.32). It is apparent that the vibration levels measured on a pump are very dependent upon operating conditions. Back pressure, suction pressure, fluid temperature, and speed should be monitored.

A common problem with vertical pumps is termed rocking mode resonance. It occurs when the first natural frequency of the pump structure matches the operating speed of the pump. The result is high vibration levels at the operating speed of the pump. The problem can be confirmed with a resonance test.

Figure 4.32. Pump Flow versus Head Curve.

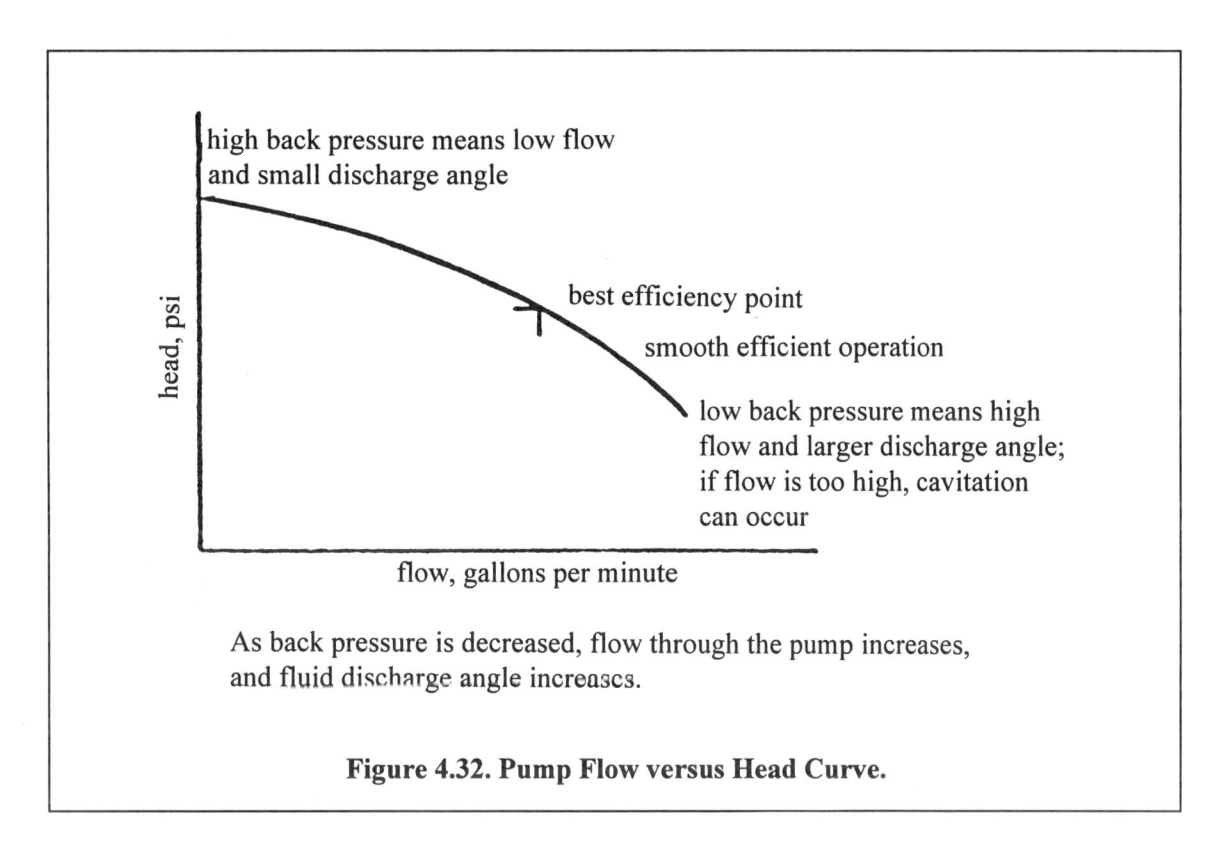

Case history of recirculation. This case history is an excellent example of the problems that occur when a pump operates against too much back pressure [4.4]. A fixed speed pump has only one back pressure, for which the flow angle of the fluid coming off the impeller matches the diffuser angle. Operation at any other point can result in inefficient operation and excessive vibration.

Antifriction bearings were failing at six-week intervals on a horizontal split case pump with a capacity of 2,400 gallons per minute (gpm) at 300 feet of total developed head. It could be seen that the rotor was moving in the axial direction at a low frequency. A head curve for the pump was requested to determine whether or not the pump was operating relative to its best efficiency point. The discharge pressure from a gauge showed that the pump was being operated at a very low flow rate, far to the left side of the pump curve. The stamp on the by-pass orifice indicated that the opening was two inches. The pump design required a three-inch orifice to assure the correct minimum flow. The recommendation was to replace the orifice. When it was removed, the hole was found to be only one inch in diameter.

This horizontal split case pump was filling a tank several floors above. When the fluid in the tank reached a predetermined level, a control valve closed. As a result the only outlet for the pump was the recirculation line. Because the orifice in the recirculation line was too small, the pump was operating against too much head.

Pumps that are forced to operate at drastically reduced flow rates build up pressure on one side of the rotor, then on the other side, due to recirculation. The result is slowly-oscillating axial forces and vibration (Figure 4.33) that can cause rapid failure of antifriction bearings that are not designed to survive the extra axial loading. All pumps with axial shuttling of the rotor should be examined to determine if they are operating against excessive back pressure.

Case history of cavitation. The pump in this case was operating against insufficient back pressure [4.4]. The result is that the pump operated in a runout condition with the fluid cavitating.

During baseline vibration monitoring, high vibration levels were discovered on the circulating water pumps at a utility. High levels were detected in both the horizontal direction on the inboard motor bearing and in the axial direction on the outboard motor bearing. The broad-band spectrum contained no mechanically-related identifiable frequencies. A spectrum of the vibration level on the inboard motor bearing is shown in Figure 4.34.

It was discovered that some of the pumps were operating against only ten feet of back pressure. A copy of the head capacity of the pump was obtained. The design capacity of the pump was 156,000 gpm at 38 feet of head. The head flow curve ended at 15 feet of back pressure, indicating that operation with only ten feet of back pressure had not even been considered by the manufacturer. An estimated flow of 200,000 gpm was obtained by projecting the head capacity curve to the ten-foot discharge-pressure level. To verify this condition, the discharge valves on the condenser were partially closed to increase the back pressure to a level closer to the design point. When the valves were partially closed, the vibration decreased to an acceptable level.

A circulating water pump removed for repair was found to have serious damage to the suction bell. It was concluded that the damage resulted from cavitation. To verify the cavitation theory a camera was installed in the suction bell of the pump to determine if cavitation was occurring. Results left little doubt that cavitation was indeed the problem. Partial closure of the condenser discharge valves dramatically reduced the cavitation.

The low discharge was caused when only one pump operated in the header instead of two. This condition occurred when the cooling-water temperature was low enough so that one pump could supply enough water to the condenser to satisfy the back pressure requirements of the turbine. Unfortunately, the one pump in operation experienced cavitation.

Figure 4.33. Recirculation.

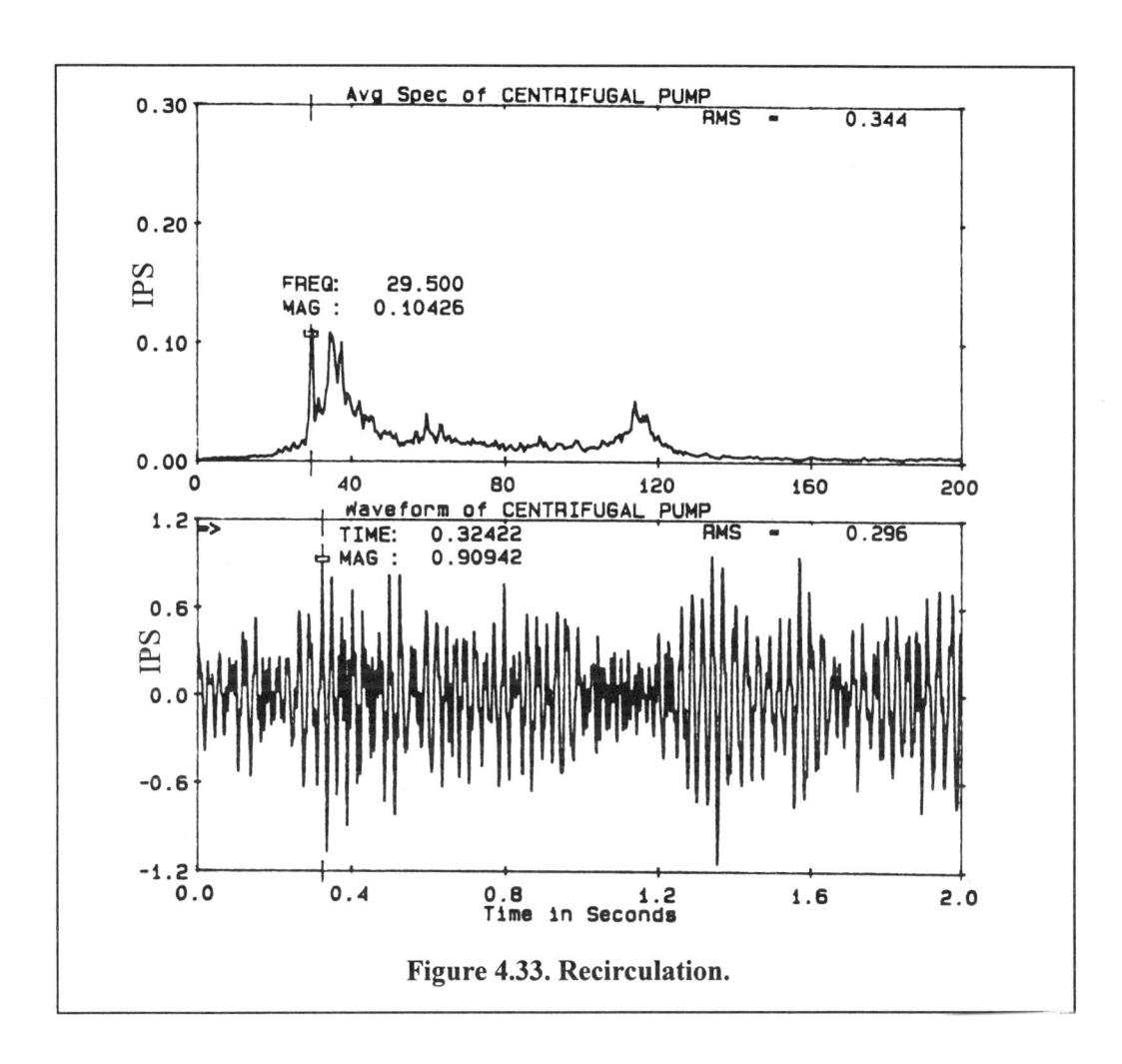

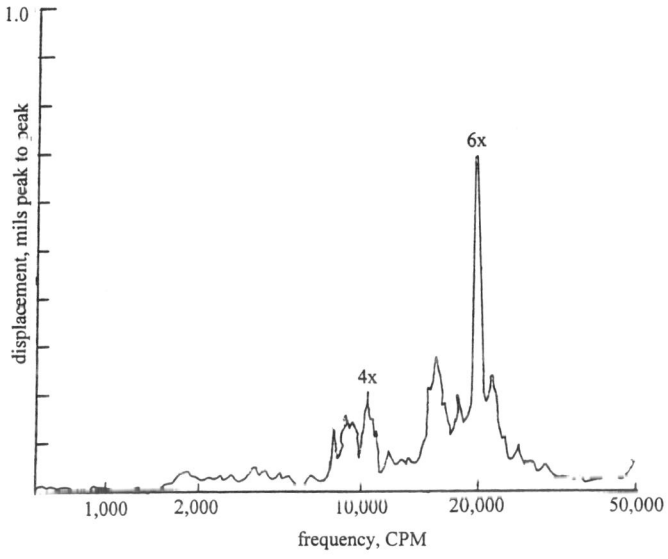

Figure 4.34. Cavitation.

Fans

Many centrifugal fans have a volute or scroll-type casing, in which the flow enters axially and leaves tangentially. Blading may be fixed or adjustable (sometimes during operation). A typical fan performance characteristic is given in Figure 4.35. The basic curve is fan pressure versus flow through the systems; head, or pressure, varies as the square of the flow. The fan will operate satisfactorily at the intersection of the system characteristics and the fan pressure characteristic. The system characteristic can be changed with an outlet damper control. Variable vane, pitch, and speed controls alter the fan characteristics. Characteristics of fans mounted in series and parallel must be considered as a system.

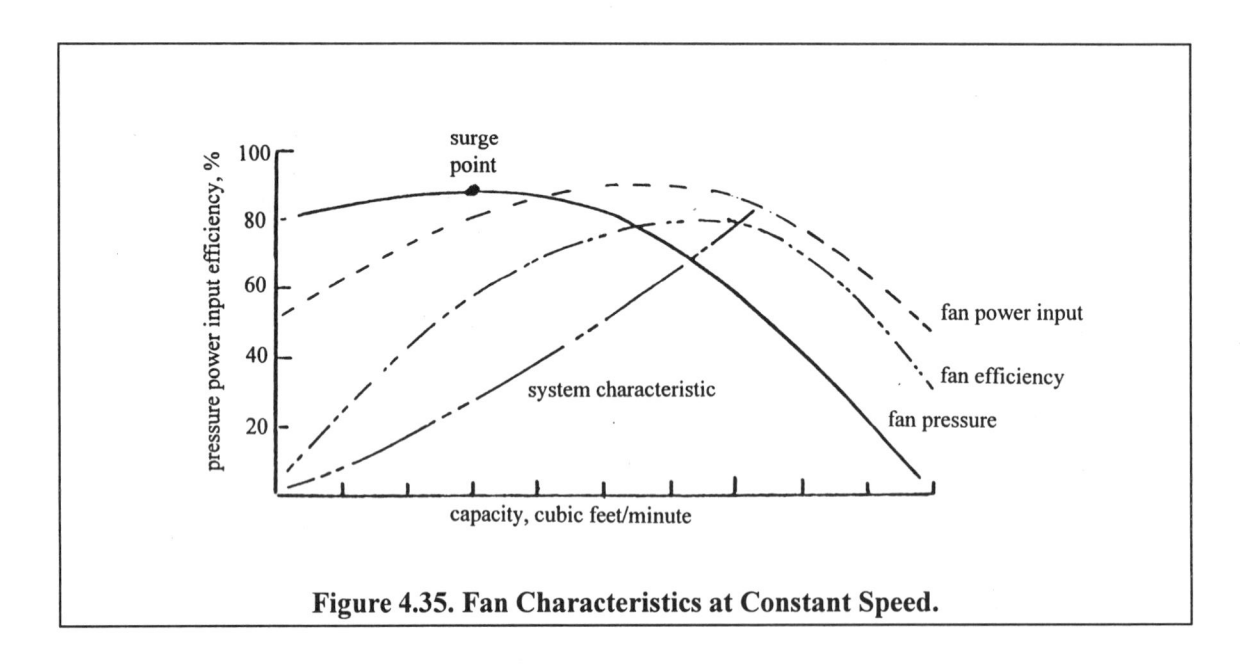

Figure 4.35. Fan Characteristics at Constant Speed.

Example 4.5: Fan drive belt.

A fan is belt-driven by a four-pole motor at 1,779 RPM with a nine-inch pulley. If the fan has an 11.75 inch-pulley attached, what is the fan speed?

fan speed = [(1,779 RPM) (9 in.)]/11.75 in. = 1,362.6 RPM

> **Example 4.6:** Fan blade-pass frequency.
> What is the blade-pass frequency for an ID fan that operates at 896 RPM and has 12 blades?
>
> BPF = (896 RPM) (12 blades) = 10,752 CPM
> = 179.2 Hz

To assure stable operation the slopes of the pressure-flow curves of the fan and system should be opposite in sign (see Figure 4.35). When the slopes of the fan and system characteristics are opposite in sign, any system disturbance that tends to produce a temporary decrease in flow is nullified by the increase in fan pressure. The condition that accompanies unsteady flow is pulsation, which occurs when the operating point of the fan is to the left of the maximum pressure on the fan curve. This is termed the surge point. Inlet dampers can usuually be used to position fan operation to the right of the surge point.

Flow separation in the blade passages of the impeller can cause unsteady flow and vibration. At low capacities, blow back or puffing can occur; that is, air puffs in and out of the inlet. Acoustic resonance occurs when a fan vane-pass frequency matches the acoustic natural frequency of the air in the duct work. Fans are subject to critical speed and structural resonance problems because they are mounted on skids, isolators, and flexible frames. Table 4.7 lists a number of common fan faults.

Table 4.7. Fan Faults.

Mass unbalance (Figure 4.36)	Isolator problems
Misalignment	Oil whirl
Critical speeds	Rolling element bearings
Resonance	Soft foot
Looseness (Figure 4.37)	Impeller eccentricity
Aerodynamic problems (Figure 4.38)	Belts and pulleys

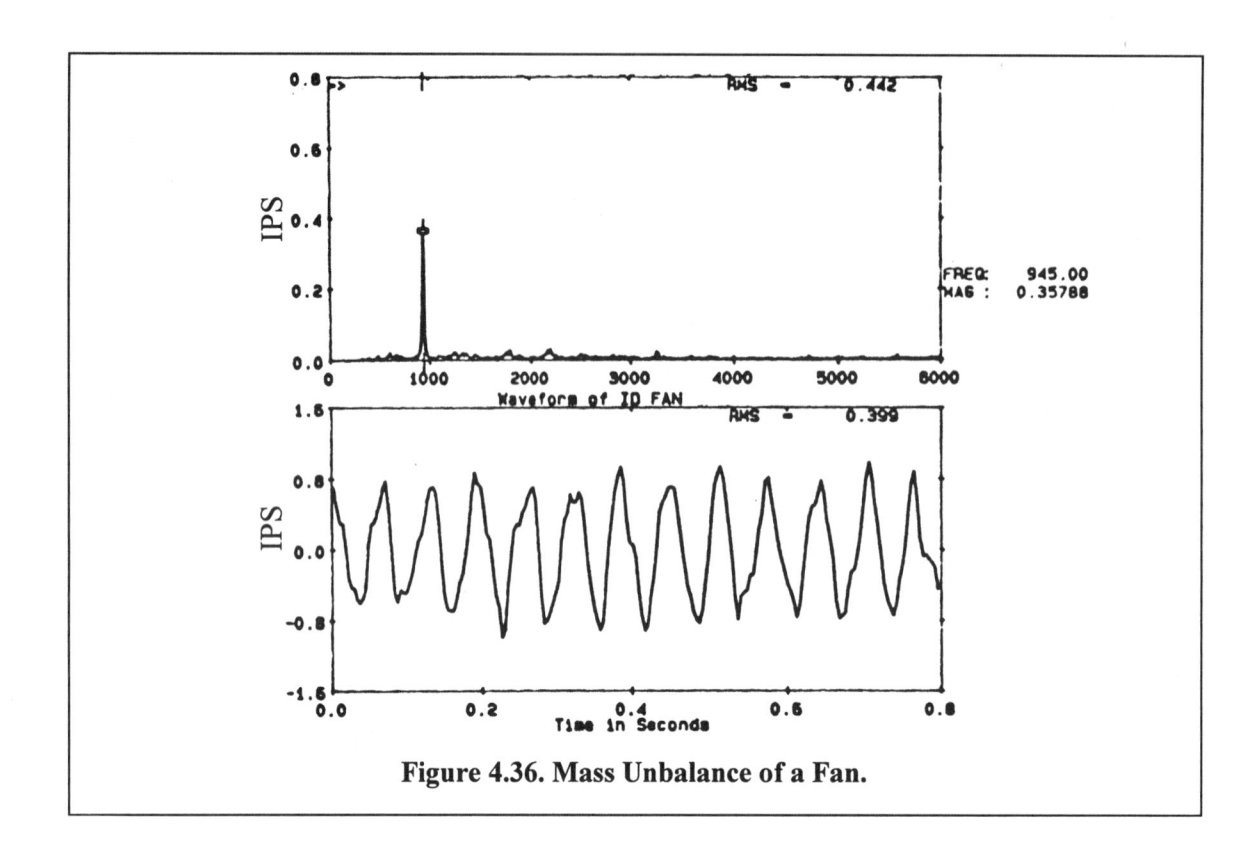

Figure 4.36. Mass Unbalance of a Fan.

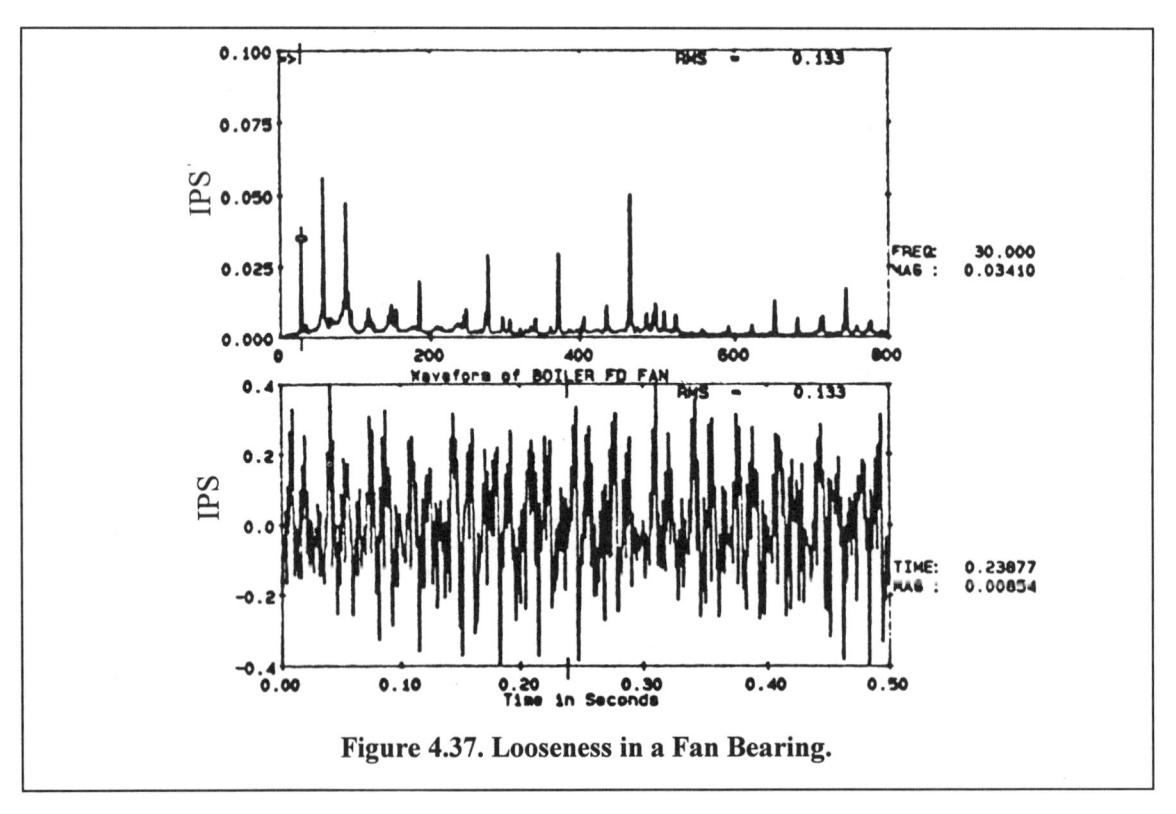

Figure 4.37. Looseness in a Fan Bearing.

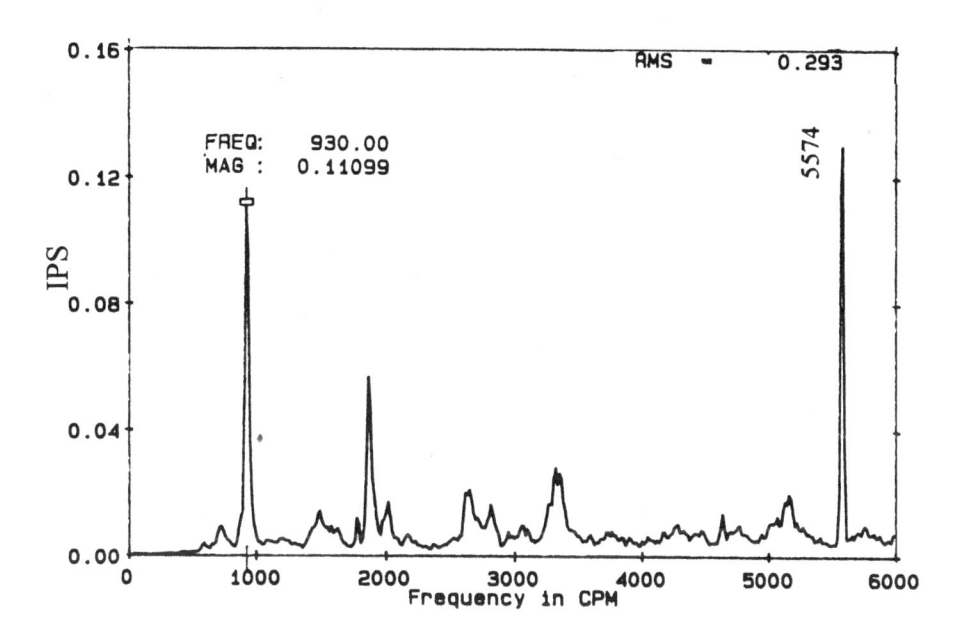

Figure 4.38. Aerodynamically-Induced Vibration in a Fan
due to Improper Positioning of Damper (Fan).

Compressors

Because of the pressures involved, most centrifugal compressors have massive casings and small lightweight rotors that make seismic vibration measurements difficult. In fact, vibration measured on the casing is severely attenuated as a result of the massive casing and the fluid-film bearings. For this reason permanently-mounted proximity probes that measure relative rotor vibration are used to analyze rotor-bearing problems.

Compressor faults are similar to those encountered in steam turbines and pumps and occur subsynchronous to operating speed, at operating speed, or as multiples of operating speed. Compressors have a minimum flow point termed the surge limit. The operation of the machine is unstable below the surge limit, which is a function of compressor type, gas properties, inlet temperature, blade angle, and speed.

Summary of Fault Diagnosis

- In general, vibration frequencies are used to determine the location of faults in a machine.
- Fault diagnosis is principally conducted in the spectrum; however, the time waveform, orbit, and phase analysis provide additional information for in-depth analysis.

- Spectrum analysis includes identification of orders of shaft speed; harmonics of gear, bearing, and vane-pass frequencies; and nonsynchronous frequencies such as bearing frequencies, beat frequencies, natural frequencies, sidebands, center frequencies, and difference frequencies.
- The spectral frequency axis (horizontal axis) can be expressed in terms of CPM, Hz, or orders.
- The spectral amplitude axis (vertical axis) can be expressed in rms, peak, or peak to peak.
- The vertical axis of the time waveform is expressed in peak units.
- Machine faults that show up at operating speed or its orders include critical speeds, mass unbalance, misalignment, rotor bow, excessive bearing clearance or wear, structural resonance, looseness, eccentricity, coupling lockup, and distortion.
- Mass unbalance occurs at the operating-speed frequency.
- Critical speeds arise when operating speed, or any of its orders containing energy, is close to or equal to a natural frequency.
- Misalignment can show up at operating speed (1x), two times operating speed (2x), or three times operating speed (3x), depending on the nature of the misalignment and the design of the shaft, couplings, and bearings.
- Shaft bow may significantly reduce vibration at a speed at which excitation is equal to and out-of-phase with mass unbalance.
- Excessive clearance and/or wear in fluid-film bearings will cause vibration similar to mass unbalance.
- Structural resonance amplifies vibration.
- Looseness appears in the spectrum at operating speed and its orders. Fractions (e.g., 1/2x, 1/3x) may also appear.
- Rolling element bearing defects occur at bearing frequencies and their harmonics. Sidebands of operating speed, fundamental train frequency, and ball spin frequency also occur, depending on the severity of the defect.
- HFD methods are used to detect pulses in machine systems.
- Gear-mesh faults arise in the spectrum at gear mesh and it harmonics. Sidebands occur as the condition deteriorates.
- Broken, cracked, or chipped gear teeth are identified as pulses in the time waveform [1].

- Eccentric gears are identified as gear mesh and sidebands at a frequency of the eccentric speed.
- Electrical problems on electric motors are identified in the spectrum as sidebands of the number of poles multiplied by slip frequency and two times lines frequency and its harmonics.
- Broken rotor bars generate sidebands at the number of poles multiplied by slip frequency at operating-speed vibration and its orders.
- Stator problems and air-gap variation arise at two times line frequency and its harmonics.
- Common problems related to pumps result from improper flow in the system, including re-circulation (high head) and cavitation (low head).
- Pump vane-pass frequencies occur if internal clearances are not set correctly.
- Fans may exhibit blade-passing frequency if aerodynamic problems occur in the duct, fan, or damper design.

References

4.1. Taylor, James I., *The Vibration Handbook*, Vibration Consultants, Inc., Tampa, FL (1994).

4.2. Shigley, Joseph E., *Mechanical Engineering Design*, McGraw-Hill Book Co., NY (1963).

4.3. Campbell, W.R., "Alternating Current Electric Motor Problems: Part 2. Electromagnetic Problems," *Vibrations*, *1* (3), p 12 (Dec 1985).

4.4. Baxter, Nelson L. *Machinery Vibration Analysis III: Part 2*, Vibration Institute, Willowbrook, IL (1995).

CHAPTER V
MACHINE CONDITION EVALUATION

Amplitudes determine a machine's condition.

The transducers and electronics now used in monitoring systems provide data that are evaluated on the basis of criteria and limits to ascertain machine condition. Computerized monitoring systems and electronic data collectors can evaluate data on the basis of overall levels of measures. Overall levels of measures are typically judged in terms of limits; e.g., acceptance of new and repaired equipment, normal, surveillance, and shutdown. These levels are compared for some period of time in order to establish trends. The levels of measures can be expressed as peak or peak-to-peak overall vibration; rotor position, peak frequency component of vibration, and overall or band root-mean-square (rms) vibration. The measure used should be based on the sensitivity of the machine — that is, the greatest change in magnitude of a measure such as velocity should be obtained for a known change in machine condition. Velocity is commonly used to measure bearing pedestal vibration because it contains both displacement and frequency and is thus a measure of fatigue. In addition, velocity is dominant in the frequency range from 10 Hz to 1,000 Hz at which most machine vibration occurs. Displacement is a good measure for machines with fluid-film bearings because it directly determines the amount of clearance in the bearing used by the vibration.

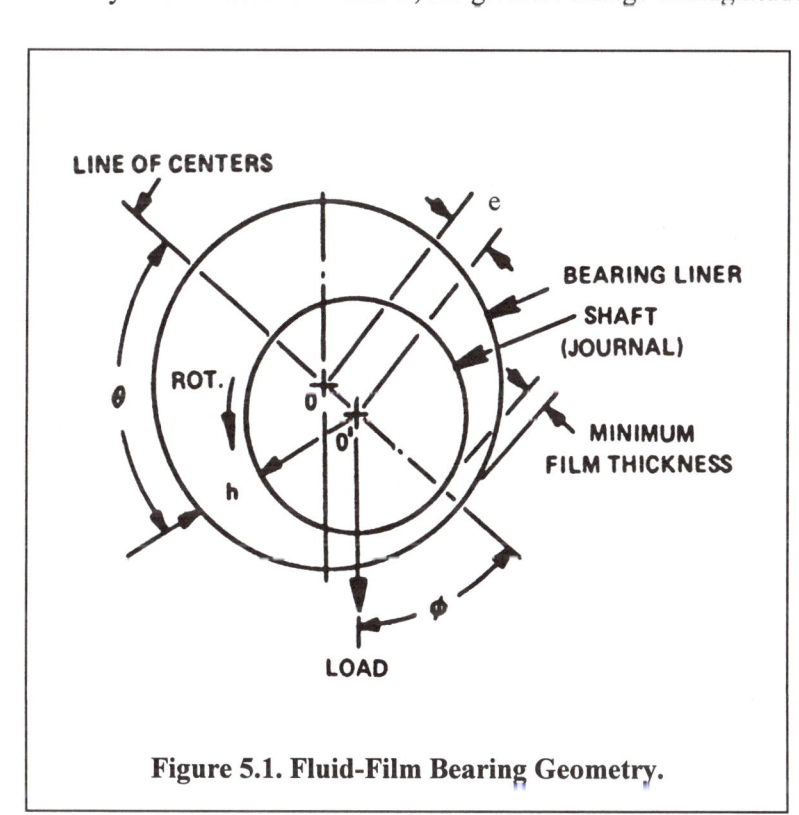

Figure 5.1. Fluid-Film Bearing Geometry.

Spectra are used when detail is required, mostly for surveillance. Each line of a spectrum is compared, either by computer or manually, to some standard or to baseline data. Automatic monitoring systems can compare the shapes of spectra point by point and overall measures to baselines or standard data.

One or more vibration measures (relative displacement, velocity, or acceleration) is monitored, depending on the design of the machine. Design factors and operational characteristics that influence the condition of a machine include speed and fatigue strength of the rotor. Bearing characteristics such as clearance are also important (Figure 5.1). The eccentricity ratio determines where the journal operates in the bearing. (The eccentricity ratio is the ratio of the eccentricity, which is the distance between the journal centerline and the bearing centerline, to the radial clearance; see Figure 5.1.) Loads on the machine must also be considered. The diversity of machine design, installation, and operational conditions has made impossible the development of absolute standards, levels, and guidelines that can be used in conjunction with monitoring systems to protect machines. Therefore, even though systems that monitor machine condition can gather accurate data very quickly, these data are of value for comparison and interpretation only if criteria and limits have been developed for a class of machines or on an individual machine basis during operation. General guidelines are available, however, that can be used to develop criteria and limits [5.1]. This chapter deals with the guidelines and techniques now available for developing vibration criteria and limits for specific machines.

Guidelines for acceptable vibration levels are based on shaft or casing measurements. Shaft vibration is used to assess the condition of a machine with large relative motions in the bearings and a high ratio of casing weight to rotor weight. Included are machines with fluid-film bearings; exceptions are the centrifugal pump and some generators. Casing and bearing cap vibrations are used in a condition monitoring program to evaluate machines with stiff bearings. Both rolling element and fluid-film bearings can be stiff or rigid, however, and their flexibility with respect to the rest of the system is important.

Shaft Vibration

Shaft vibration is measured with proximity probes at or as close as possible to the bearings. Such measurements are useful if relative motion can adequately provide sensitivity. A machine with a stiff bearing is not sufficiently sensitive.

If two probes are used at the bearing, an orbit of its motion, as well as the position of the journal inside the bearing, can be obtained while the machine is operating. Proximity probe measurements establish the equilibrium position of the journal. The dynamic signal provides the position at any instant. From this information an accurate assessment of bearing condition can be made directly from measurements. A guideline for evaluating shaft vibration based on bearing measurements is shown in Figure 5.2. Normal, surveillance, and shutdown limits are given. In addition, evaluation levels of machine condition related to shaft motion have been published by the International Standards Organization [5.3]

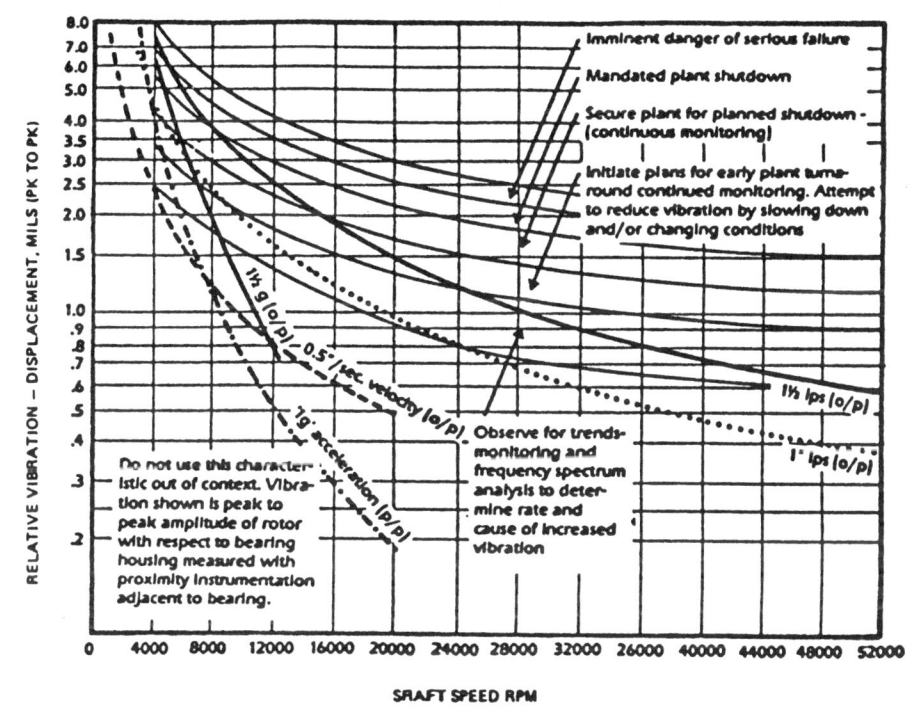

Figure 5.2. Dresser-Clark Chart for Measurement of Shaft Vibration on Turbomachinery with Proximity Probes [5.2].

Bearing Vibration

A widely used method for evaluating the vibration of the journal in a fluid-film bearing is to compare the relative vibration of the rotor to the clearance in the bearing. Table 5.1 relates journal bearing clearance, rotor speed, and relative vibration to recommended maintenance actions. The ratio R/C of the measured relative vibration R (in mils peak to peak) to the diametral clearance C (in mils peak to peak) of the bearing is calculated and identified in Table 5.1 according to machine speed. Diametral clearance is the difference between the bearing diameter and the journal diameter. Figure 5.3 contains data from an axial compressor bearing. The machine speed is 1,437 RPM, and the diametral clearance of the journal is 12 mils. Peak-to-peak vibration, measured on the time waveform, is 5.63 mils. The R/C value is 5.63/12, or 0.47. At speeds lower than 3,600 RPM, the data from Table 5.1 indicate surveillance of the machine.

Table 5.1. Evaluation of Rotor/Bearing Vibration.

Maintenance	Allowable R/C	
	3,600 RPM	**10,000 RPM**
Normal	0.3	0.2
Surveillance	0.3-0.5	0.2-0.4
Shut down at next convenient time	0.5	0.4
Shut down immediately	0.7	0.6

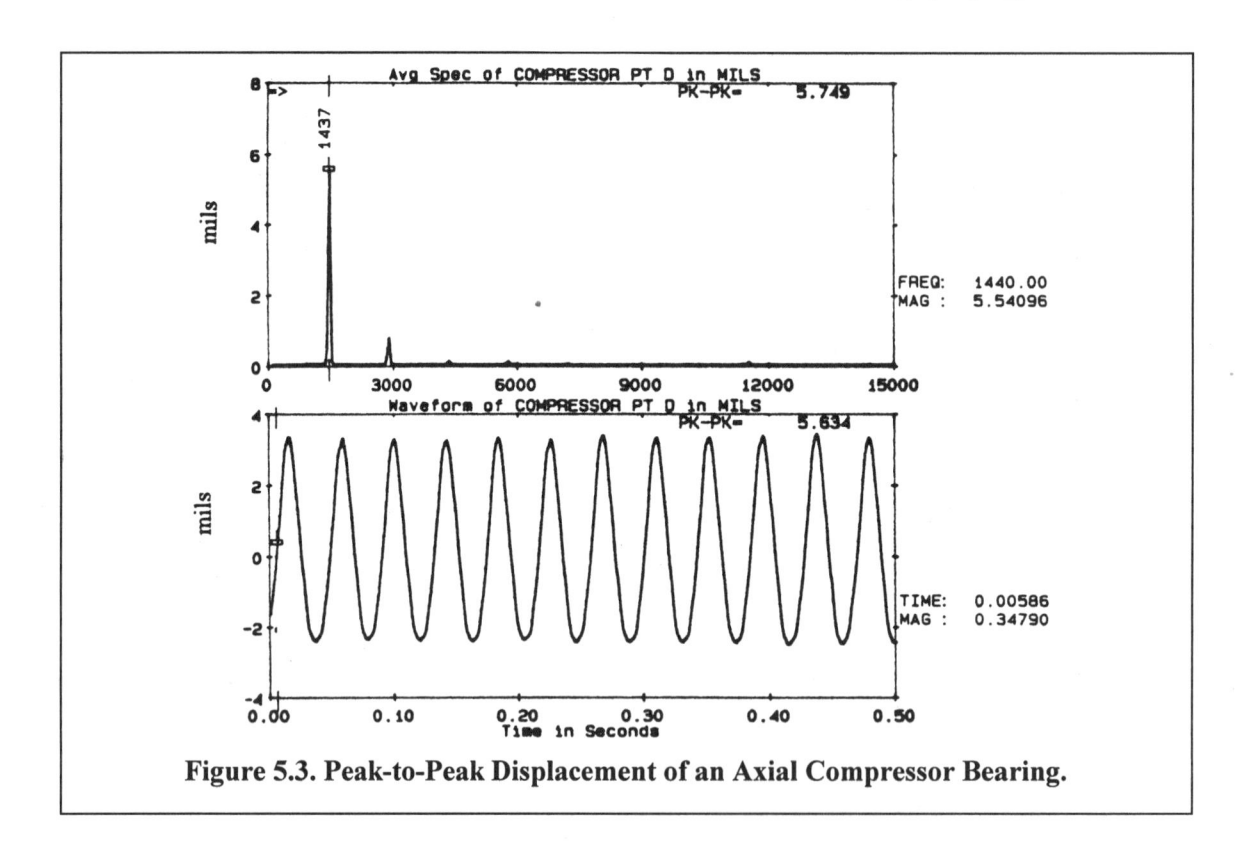

Figure 5.3. Peak-to-Peak Displacement of an Axial Compressor Bearing.

Example 5.1: Determination of Condition and Maintenance Action (Table 5.1).

Determine the condition and maintenance action required, according to Table 5.1, of data from the motor shown in Figure 4.8. The speed is close to 3,600 RPM and the diametral clearance is 9.0 mils. The maximum displacement is 4.3 mils peak to peak. Therefore,

$$R/C = 4.3/9 = 0.48$$

This unit is in marginal condition as a result of excessive mass unbalance. It is operating close to the shut-down level and should be balanced at the next convenient opportunity. Depending on the date of bearing installation, inspection of the bearing during shutdown would be prudent.

Casing Vibration

In some machines large vibratory forces are transmitted through the bearings to the casing. Vibration measurements should be made at the bearing cap on the casing, as close to the bearing as possible. Velocity transducers or accelerometers should be used to measure the vibration. The type of measurement made is dependent on the design and operating conditions of the machine. The theoretical basis of vibration severity limits and the relationship between shaft and casing measurements has been examined [5.4].

Most tables and charts available for assessing vibration limits on bearing caps [5.5] apply to general-purpose machines. The limits are based on overall peak or rms measurements and were developed for the once-per-revolution-component of vibration.

A modified Blake chart shown in Figure 5.4 relates to overall peak vibration adjusted with a service factor to obtain effective displacement, velocity, or acceleration. The peak vibration at the bearing cap must be measured using the time waveform or a peak detecting circuit. Note that the horizontal and vertical axes are log scales. They are used to compress the scales so that the entire range of data can be applied to a single chart. In this case, data from 0.01 IPS to 10 IPS can be evaluated with adequate dynamic range. The speed and type of machine must be noted. The level of vibration is applied to Figure 5.4 at the appropriate speed to obtain the machine condition. For example, a turbine generator with a measured vibration of 0.15 IPS falls in the "Some Fault" region of the chart (Figure 5.4). Maintenance actions recommended in Table 5.2 for values of peak or rms velocity are based on extensive field data. Figure 5.5 contains data taken from a single-stage centrifugal pump operating at 1,770 RPM. The measured peak vibration is 1.76 IPS (from the time waveform in Figure 5.5).

Table 5.2. Vibration Guidelines for Condition Evaluation.*

Condition	Limits	
	rms velocity	peak velocity
Acceptance of new or repaired equipment	<0.08	<0.16
Unrestricted operation — normal	<0.12	<0.24
Surveillance	0.12-0.28	0.24-0.7
Unsuitable for operation	>0.28	>0.7

*These values should be adjusted to reflect the condition of the machine.
Service factors may be necessary for some special equipment, depending on design, speed, and/or process.

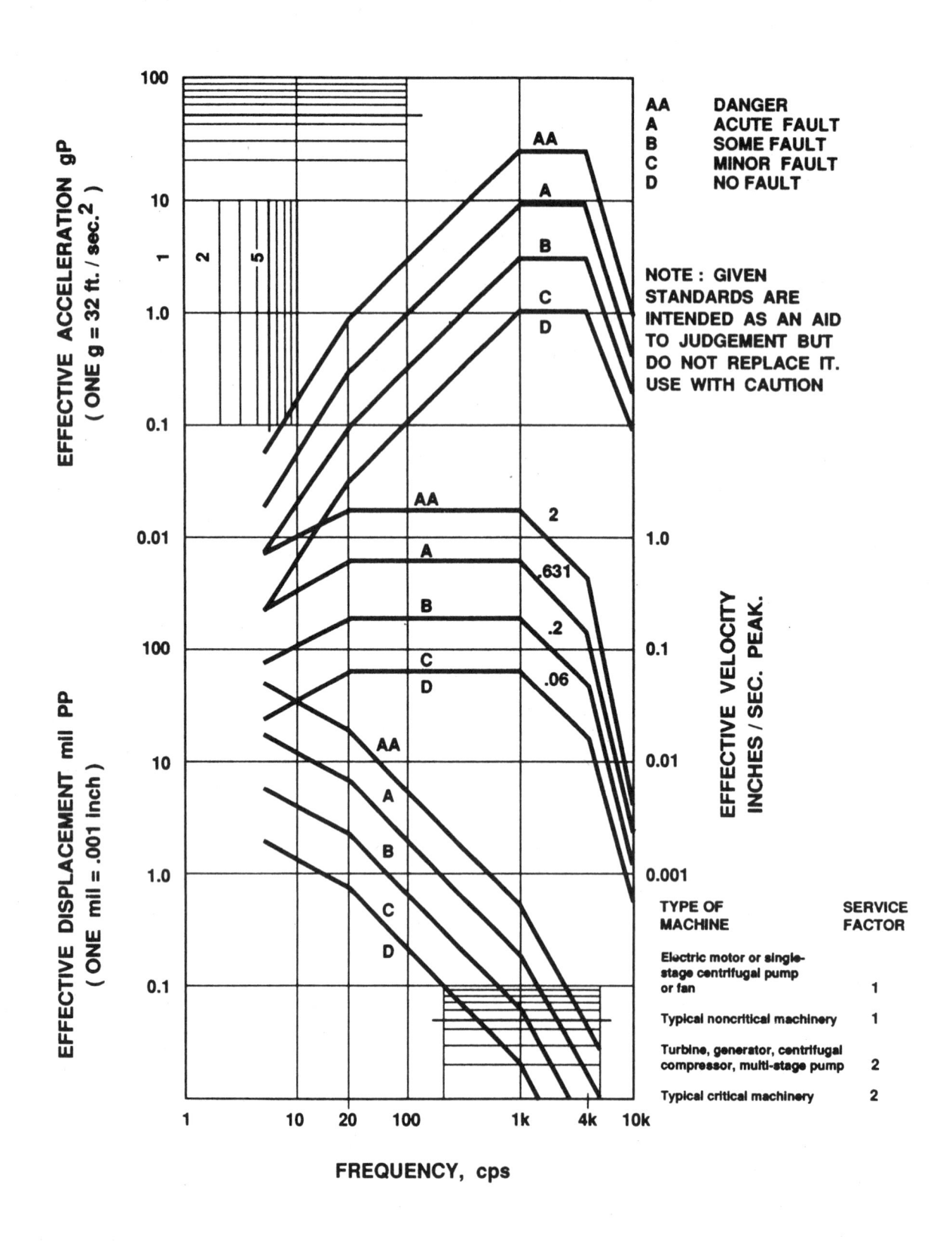

Figure 5.4. Modified Blake Chart for Peak Bearing Cap Vibration Limits [5.5].

Criteria and limits for vibration levels measured on the casing have not been formalized for specific machines. Overall guidelines that utilize a service factor have been established for evaluating once-per-revolution faults [5.5]. These guidelines are based on peak vibration velocity measured on the casing. The guidelines have been customized by comparing measured data from specific machines with known problems to the levels given in the guidelines. In subsequent measurements the effective vibration evaluated in the guidelines is obtained by multiplying the measured vibration by the service factor. For example, if 0.3 IPS were found to be satisfactory for a rotary blower, a service factor of 0.2/0.3 = 0.667 would be established for the Blake chart. Service factors cannot be established on the basis of one measurement. A statistical sample of the relationship between machine condition and measured vibration is required.

The acceptance levels shown in Table 5.3 [5.6] include a wide variety of equipment and are applied to new and repaired equipment. The data were developed from experience with vibration on bearing caps measured as rms vibration velocity [5.6, 5.7].

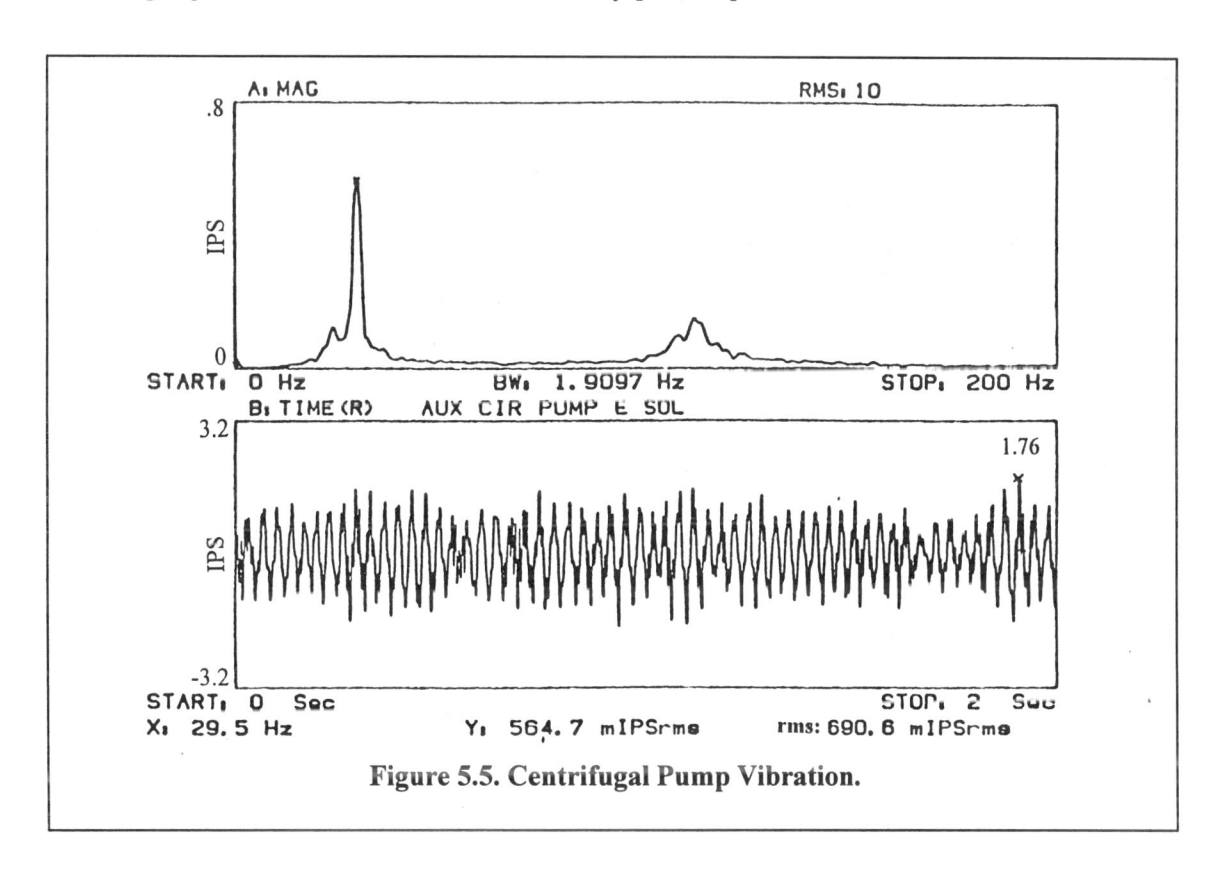

Figure 5.5. Centrifugal Pump Vibration.

The limits are based on the size of the machine. If the data collector provides rms or derived-peak (1.414 x rms) data, the ISO chart [5.6] or Table 5.2 should be used for evaluation.

> **Example 5.2:** Determination of Condition and Maintenance Action (Table 5.2).
>
> Determine the condition and maintenance action required for data from the lube oil pump shown in Figure 4.14. The pump operates at 3,488 RPM and has an rms of 0.123 IPS. Therefore, the level must be compared to an rms chart (Table 5.2) According to Table 5.2 the unit is in the surveillance range and should be monitored for changes in vibration level.

Summary of Machine Condition Evaluation

rms velocity ranges of vibration severity in./sec.	vibration severity* for separate classes of machines			
	Class I	Class II	Class III	Class IV
0.01				
0.02	A			
0.03		A		
0.04	B		A	
0.07		B		A
0.11	C		B	
0.18		C		B
0.28	D		C	
0.44		D		C
0.71			D	
1.10				D
1.77				

*The letters A, B, C, and D represent machine vibration quality grades, ranging from good (A) to unacceptable (D).

Class I. Individual components, integrally connected with the complete machine in its normal operating conditions (i.e., electric motors up to 15 kilowatts, 20 HP).
Class II. Medium-sized machines (i.e. 15- to 75-kilowatt electric motors and 300-kilowatt engines on special foundations).
Class III. Large prime movers mounted on heavy, rigid foundations.
Class IV. Large prime movers mounted on relatively soft, lightweight structures.

Table 5.3. ISO 2372 1974 [5.6].

- In general, the severity of machine condition is assessed by using the amplitude of vibration.
- As a result of variation in design and low signal strengths, rolling element bearings and gears require the evaluation of amplitudes and frequencies.
- Principal measures for bearing cap (casing) vibrations are peak or rms velocity and acceleration.
- All bearing cap measures should be stated in either rms or peak form and should not be mixed.

- Shaft vibration severity is evaluated using relative displacement peak to peak, bearing diametral clearance, and rotor speed.
- Shaft vibration is the preferred basis for evaluating machines with a large casing-to rotor weight ratio. For casing measurements, a significant service factor (3-5) must be used.

References

5.1. Maedel, P.H., Jr., "Vibration Standards and Test Codes," *Shock and Vibration Handbook*, 4th edition, C.M. Harris, ed, McGraw-Hill, NY (1996).

5.2. Jackson, C., *The Practical Vibration Primer*, Gulf Publishing, p 46 (1979) (out of print).

5.3. ISO 7919, 1986, "Mechanical Vibrations of Non-Reciprocating Machines – Measurements on Rotating Shafts and Evaluation," International Standards Organization, Geneva, Switzerland*.

5.4. Maxwell, A.S., "Some Considerations in Adopting Machinery Vibration Standards," *Proceedings*, Machinery Vibration Monitoring and Analysis Meeting, Vibration Institute, Willowbrook, IL, pp 97-107 (1982).

5.5. Blake, M. and Mitchell, W., *Vibration and Acoustic Measurement Handbook*, Sparten Books, NY (1972) (out of print).

5.6. ISO 2372, 1974, "Mechanical Vibrations of Machines with Operating Speeds from 10 to 200 RPM – Basis for Specifying Evaluation Standards," International Standards Organization, Geneva, Switzerland*.

5.7. ANSI S2.41, 1985 (R 1990), "Mechanical Vibration of Large Rotating Machines with Speed Range from 10 to 200 Rev/s – Measurement and Evaluation of Vibration Severity in Situ" American National Standards Institute, NY**.

* ISO Standards can be obtained from the Director of Publications, American National Standards Institute, NY, NY 10005-3993.
** ANSI Standards can be obtained from the Acoustical Society of America, Standards and Publications Fulfillment Center, P.O. Box 1020, Sewickley, PA 15143-9998.

CHAPTER VI
MACHINE TESTING

A test is worth a hundred analyses made on paper.

Machine tests other than periodic monitoring are conducted to gain information about the design or condition of a machine. A machine is tested for various reasons: acceptance, baseline data for periodic monitoring, design verification (damping and natural frequencies), fault diagnosis, condition evaluation, and balancing.

Test Plans

It is a mistake to generate a test plan at a site on an as-needed basis. It is necessary to think through the goals of a test and related test specifications so that important data are not missed. The test plan should include a description of the machine, the test types and data to be acquired, loads, speeds, machine configurations, and process conditions. The data acquisition plan presented in Table 6.1 is tailored to the fault/condition analysis of a turbine-gear-generator unit (Figure 6.1) using casing and shaft measurements at a load of 8 Mw.

Generator Speed 1800 RPM
           Load   8   MW

RECORD NO.	MEASURE (UNITS)	1	2	3	4	5	6	7	8	PURPOSE
1	Velocity (IPS)	1X	1Y	1Z	2X	2Y	2Z	1T*	7T**	Basic Turbine Analysis
2	Velocity or Acceleration	3R	3A°	5R	5A°	6R	6A	1T	7T	Basic Gearbox Analysis
3	Velocity	7X	7Y	7Z	8X	8Y	8Z	1T	7T	Basic Generator Analysis
4	Displacement (Mils-Pk to Pk)	1V	1H	2V	2H	THRUST A	THRUST B	1T	7T	Turbine Shaft Vibrations
5	Displacement (Mils Pk to Pk)	7V	7H	8V	8H	3V	3H	1T	7T	Generator/Gearbox Shaft Analysis
6	Displacement (Mils-Pk to Pk)	6V	6H	4A°	4R°	3A°	5A°	1T	7T	Gearbox Shaft/Casing Analysis
7	Velocity	1Y	2Y	3R	6R	7Y	8Y	1T	7T	Cross Sensitivity
8	Velocity	1X	2X	3A	6A	7X	8X	1T	7T	Cross Sensitivity
9	Velocity	3Z	4Z	5Z	6Z	2Z	7Z	1T	7T	1X Phase Analysis

* 1X trigger from optical pickup or displacement probe mounted on the turbine shaft
** 1X trigger from optical pickup or displacement probe mounted on the generator shaft
° acceleration, g
* velocity, IPS

Table 6.1. Data Acquisition Plan for a Turbine Generator.

Data are acquired on an eight-channel digital tape recorder at various locations on the three machines. These data are acquired to perform time waveform, spectrum, phase, orbital, synchronous time, and cross-channel (dual) analyses. Record No. 1 involves the recording of horizontal, vertical, and axial data on the turbine governor and drive ends. Triggers are recorded on both turbine and generator shafts to permit filtering at operating speed (1x), synchronous time average, and axial phase analysis at 1x. Record No. 2 provides data for gearbox analysis, including velocity for operating speed vibrations and acceleration for gear-mesh vibrations (3,240 Hz). Record No. 3 provides basic casing data for analysis of the generator. Records No. 4 and No. 5 have to do with the acquisition of shaft vibrations on the turbine-generator and gear drive shaft for orbital analysis. Record No. 6 is concerned with casing and shaft vibrations for gearbox analysis. Records No. 7 and No. 8 are obtained for dual-channel cross-sensitivity analyses. Record No. 9 provides data for phase analysis at operating speed for the three machines.

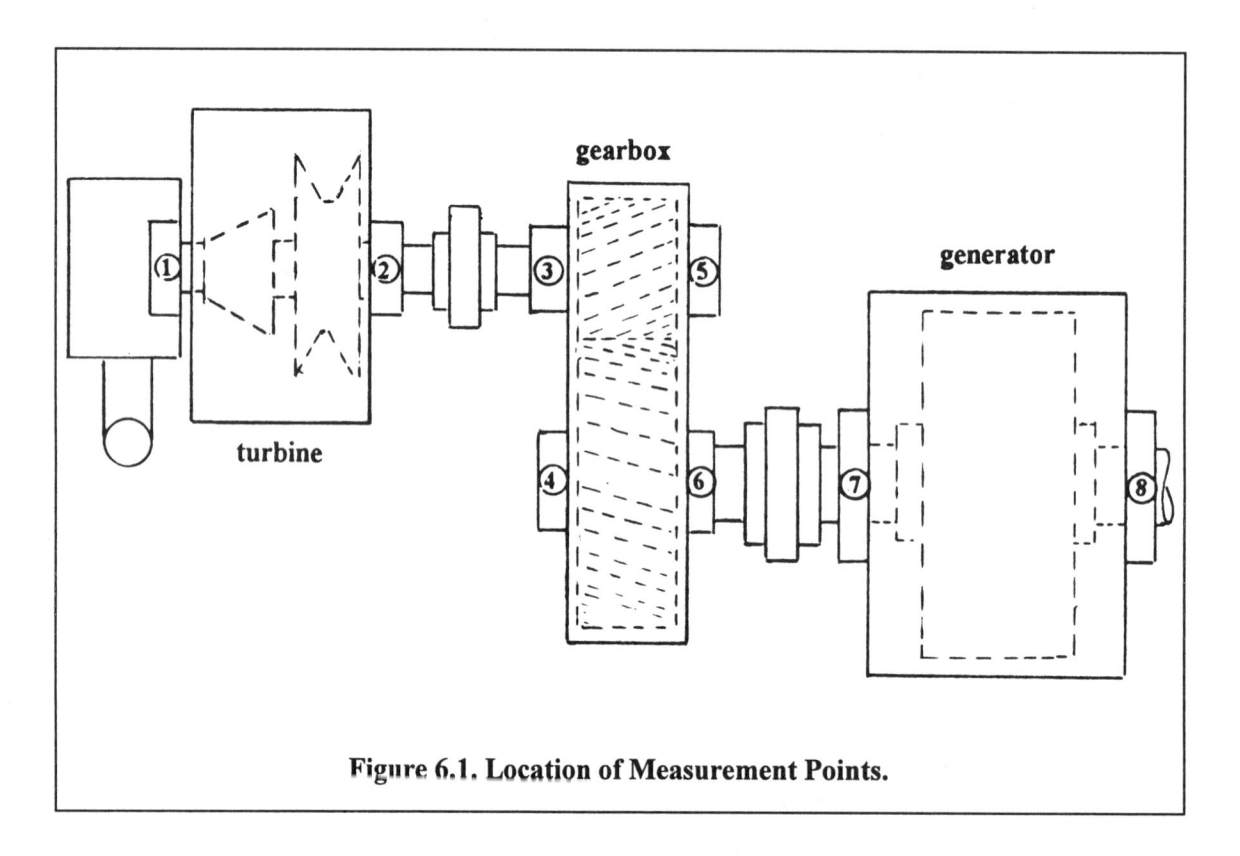

Figure 6.1. Location of Measurement Points.

If the data are recorded using a tape recorder, many different types of analyses can be performed without the machine being out of service for an extended period of time. If the vibration levels are sensitive to speed or load, extra sets of data will be required as dictated by the test plan.

A data collector can be used for acquisition of these data. However, data acquisition in the field will be more extensive and time consuming. Frequency spans, windows, and number of lines must be selected prior to data acquisition. No reprocessing of data is possible after data are acquired with a data collector. Therefore, it is recommended that the data acquisition plan be set up as a route on the data collector. If orbits, synchronous time averaging, and cross-channel analyses must be performed in the field, a two-channel data collector will be required. These considerations mean that a very detailed plan must be created when a data collector is used for data acquisition. For example, the frequencies present and the resolution required must be determined before the data are acquired.

In periodic monitoring the route that has been thought out in advance serves as the test plan. Baseline data collection requires various tests — impact, start-up, and coast-down tests — to identify natural frequencies as well as routinely collected data from the newly established route. The data acquisition plan in all cases requires a thorough description of the equipment and the location of measurement points (see Figure 6.1). Discussions of the measures and frequency spans required for data acquisition can be found in *Chapter I* and *Chapter II*. The machine configuration and process conditions are unique to the equipment being tested. The type of diagnostic test used is concerned with the goal of the plan. Operating-speed tests are conducted to obtain data for fault analysis and condition evaluation. Impact and start-up/coast-down tests are utilized to obtain natural frequencies and critical speeds. Acceptance tests are conducted to determine whether or not the new or repaired equipment meets the purchase specification. Baseline tests are used to acquire vibration data that are normal to the machine. Calibration tests are conducted for information on balance-weight sensitivity and phase lags in the machine.

Selection of Test Equipment

The test equipment required to carry out a data acquisition plan depends on the goals of the plan and the equipment available. Selection of the transducer, tape recorder, and analyzer are important. For example, if low frequencies or high temperatures are involved, special transducers may be needed. If rapid tracking for start-up and coast-down tests is needed, an FFT analyzer may not be adequate. Instead, a tracking filter will be required. Many times a data collector is used for these tests. Modern data collectors are useful in performing 95% of the work, including data storage.

If a data collector is used to store data, however, the data collected must be elaborate because reprocessing is not possible. Data that have been recorded, on the other hand, can be processed to obtain optimal information within the frequency response characteristics of the tape recorder.

Site Inspection

Site inspection and evaluation are important regardless of the type of data acquisition plan. Bolts, foundation, grouting, piping, and thermal conditions must be known. These factors often account for excessive vibration. It is necessary to eliminate nonoperating speed components of vibration by assessing the environment when the equipment is not operating and by obtaining time-averaged data.

Acceptance Tests

The acceptance test is based on a purchase specification that includes procedures, measurement locations, process conditions, measures and how they are processed, and acceptable levels of vibration. If no specification exists, a baseline test should be conducted and the data compared with general vibration standards. The baseline test should reflect the operating conditions of the machine and its environment to the best extent possible.

The purchase specification should include testing procedures as well as acceptable levels of vibration; that is, it should be similar to ISO standards. For example, ISO 2732 [6.1] contains information about equipment mounting, the measures to be used, transducer locations, and acceptance levels. A listing of a number of codes and standards for machines is also available [6.2].

Baseline Tests

Baseline tests are conducted prior to and during periodic monitoring program activity. The baseline test is used to determine the nature and level of normal vibrations of a machine. It is a known fact that different machines operate normally at different vibration levels and are often higher than general severity levels (*see Chapter V*). When baseline vibration levels change, condition can be observed and maintenance action initiated. In addition, if the vibration is initially high as a result of an installation — for example, alignment, soft foot, distortion — or design problems such as resonance or critical speeds occur, action can be planned. Records No. 1-No. 6 (see Table 6.1) would be obtained in a baseline analysis.

Resonance and Critical Speed Testing

Resonance and critical speed tests are carried out to obtain information about the dynamic characteristics of a machine and its structural support and piping. Information about resonances and critical speeds is helpful in machine diagnostics and when a machine and its associated structures must be redesigned to overcome chronic problems.

Resonances and critical speeds are frequencies that are governed by the design and operating condition of the machine. A resonance is a condition in a structure or machine in which the frequency of a vibratory force such as mass unbalance is equal to a natural frequency of the system. If the vibratory force is caused by a rotating machine, the resonance is called a critical speed. At or close to this speed the vibration is amplified. Testing techniques for natural frequencies of structures differ from machines because the machines generally have speed-dependent dynamic characteristics. Machines are tested at critical speeds to obtain the best data. Resonances are often artificially excited with hammers and shakers to obtain natural frequencies of foundations, structures, and piping.

This section is concerned with basic concepts and instrumentation used to determine the dynamic characteristics of machines and their associated structures, foundations, and piping. Testing techniques for the determination of natural frequencies are described.

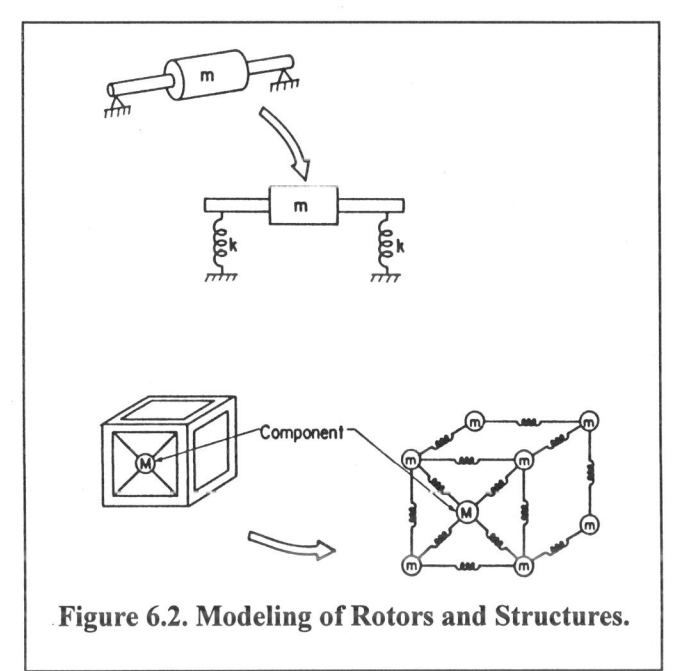

Figure 6.2. Modeling of Rotors and Structures.

Natural frequencies and mode shapes. The natural frequency of a machine or structure is governed by its design. A machine can be represented by weights connected to springs, as shown in Figure 6.2. Each machine system has a number of natural frequencies that can be excited by impact, random forces, or harmonic vibrating forces of the same frequency. In general, natural frequencies are not multiples of the first natural frequency; exceptions are rare instances involving simple components. A natural frequency becomes important in machine diagnosis when a forcing frequency occurs at or close to a natural frequency. A state of resonance then occurs that produces high vibration levels.

Close to resonance or critical speeds, vibration levels — governed by vibratory forces and damping — that might otherwise be normal are then amplified to the point that excessive vibration can structurally damage the machine— particularly the bearings.

Mode shapes of a system are associated with natural frequencies. A mode shape is defined as the deflection shape assumed by a system vibrating at a natural frequency. A mode shape does not provide information about absolute motions of a system. (Damping and vibratory forces must be known to obtain absolute motions.) Rather, it consists of deflections at selected points in the system that are determined relative to a fixed point, usually the end of a shaft. Two mode shapes for a turbine rotor are shown in Figure 6.3.

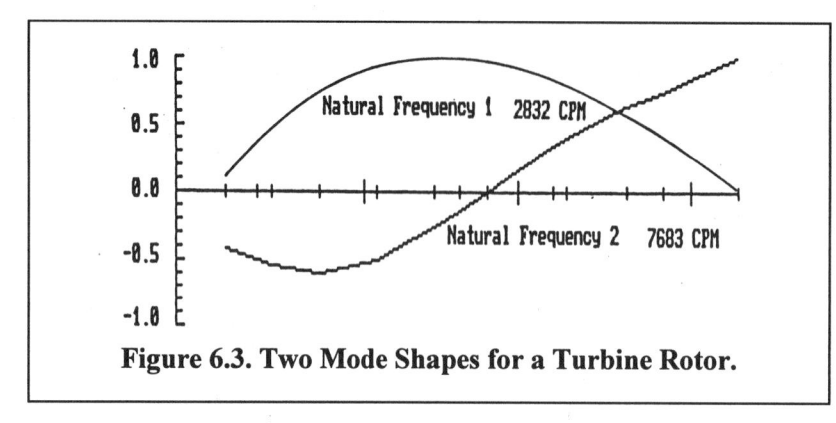

Figure 6.3. Two Mode Shapes for a Turbine Rotor.

Excitation. A machine or structure can be excited by one or more vibratory forces. The force may have a single constant frequency, as occurs with mass unbalance. Multiple frequencies occur in engines and reciprocating compressors. A variable single frequency is typical of a synchronous motor on start-up. An example of random frequencies is pump cavitation. Vibratory forces can be caused by various factors including design, installation, manufacture, and wear.

Interference diagrams. An interference diagram is used to locate resonances and critical speeds with respect to operating speed. The vertical axis (see Figure 6.4) usually contains the natural frequencies and forcing frequencies. The horizontal axis is the operating speed of the machine. The point of intersection of the one or more forcing frequencies and the natural frequency is a critical speed. Whether or not the point is actually a critical speed depends on the amount of force and damping in the system. Figure 6.4 is an interference diagram for a rotor subject to mass unbalance. A single forcing frequency

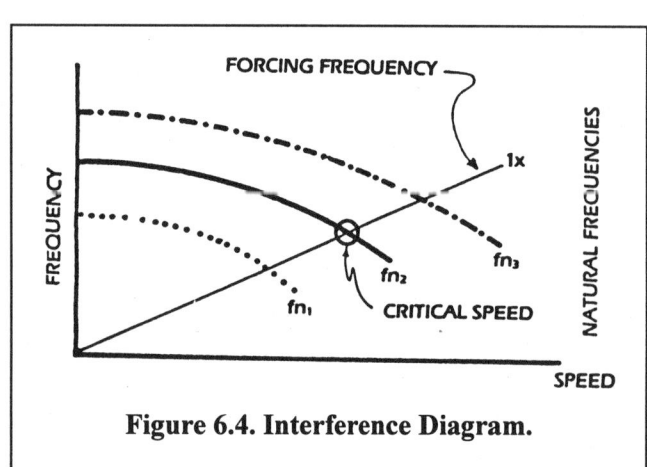

Figure 6.4. Interference Diagram.

6.6

is present. An interference diagram can be generated from computer models or test data. Computer-generated interference diagrams are often validated using test data.

Conducting a resonance test. Determine the vibrations of the structure at a number of known points during operation (Figure 6.5). These points provide guidance for impact and measurement locations.

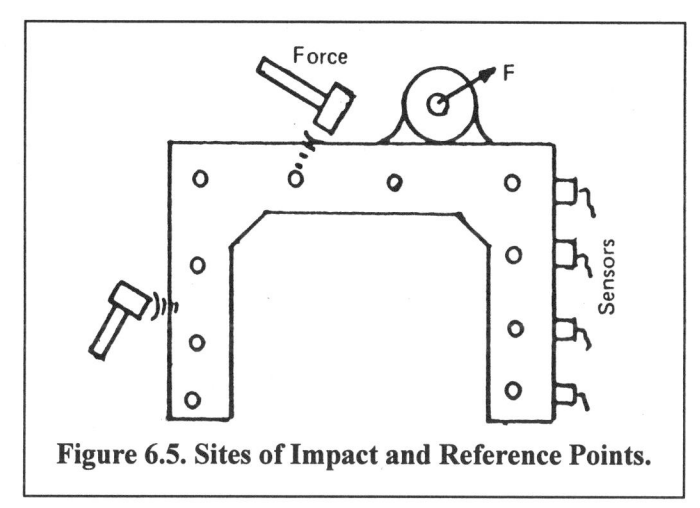

Figure 6.5. Sites of Impact and Reference Points.

Set up the data collector or analyzer for data acquisition and processing. The trigger should be set on the vibration signal — data will be acquired upon impact. The frequency span should be wide enough to view the suspected natural frequency yet provide sufficient resolution to obtain accurate natural frequencies. Use a uniform window. If averaging (multiple strikes) is used, do not strike more often than the FFT data acquisition time. For example, with an F_{max} of 100 Hz used with 400 lines the data acquisition time is 400/100, or four seconds. Only one impact should be made in four seconds. Double hits within the data acquisition time result in noisy spectra.

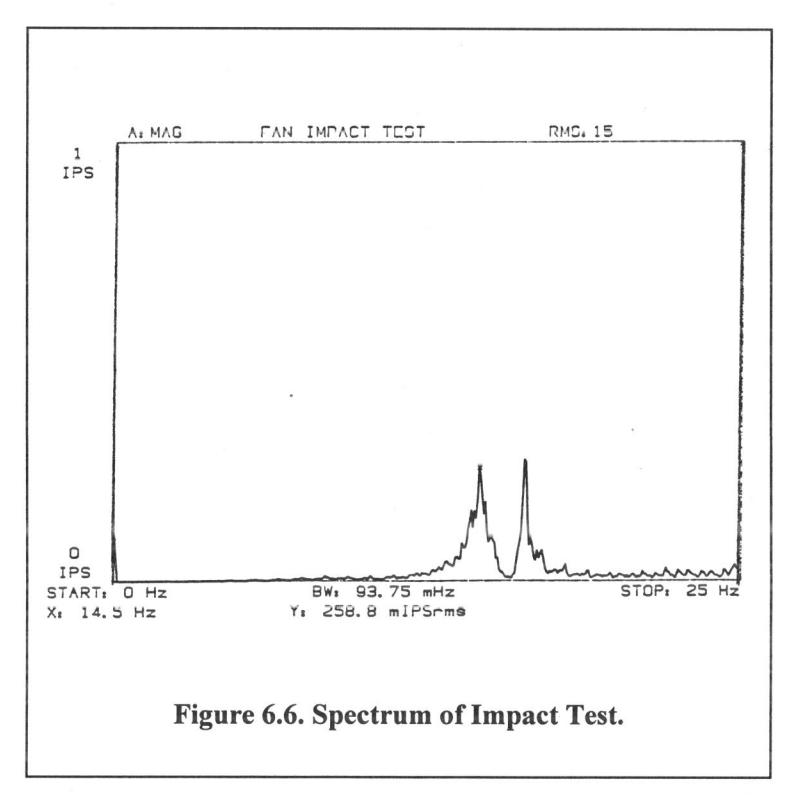

Strike the structure with a 4x4 timber, mallet or hammer with a soft head in the direction of the desired mode. If the desired mode is not known, strike the structure in several directions. The directions shown diagrammatically in Figure 6.5 vertical and horizontal usually provide useful data. Measure and record the vi-

Figure 6.6. Spectrum of Impact Test.

6.7

bration levels at a number of reference points on the structure (see Figure 6.5).

The peaks on the spectrum of vibration levels at various measurement points indicate the natural frequencies of the structure (Figure 6.6). Some natural frequencies are not seen at all measurement points. These are nodal points. Figure 6.6 is the spectrum of an impact test on the bearing pedestal of an overhung fan. The operating speed of the fan is 935 RPM. The mass unbalance-excited vibration is at 15.6 Hz, which is between the 14.6 Hz and 16.4 Hz natural frequencies but too close to both natural frequencies. The rule of thumb for spacing between forcing frequency (mass unbalance) and natural frequency is 15% (2.34 Hz in this case).

For resonance testing, the structure, piping, or machine should be as close as possible to its operating state. Parts of a machine cannot arbitrarily be removed and tested. For example, the natural frequencies of a gear not mounted on its shaft differ from those when the gear is mounted. Similarly, the natural frequencies of a machine mounted for shop testing differ from those of the machine mounted on its normal foundations.

Conducting a critical speed test. Select one or more appropriate transducers to measure the vibration. Proximity probes measure relative shaft displacement and are preferred if they are permanently installed. Otherwise, mount seismic transducers — either velocity transducers or integrated accelerometers — as close to the bearing as possible in the horizontal and vertical directions. For a permanent or temporarily-mounted trigger use a proximity probe or a magnetic pickup adjacent to an indentation or mark on the rotor. A combination of an optical pickup and reflective tape may also be used. Wire the vibration transducers and trigger to a tracking filter, tape recorder, or data collector. If the vibration transducers and trigger are permanently mounted, a coast-down test of the machine can be performed directly.

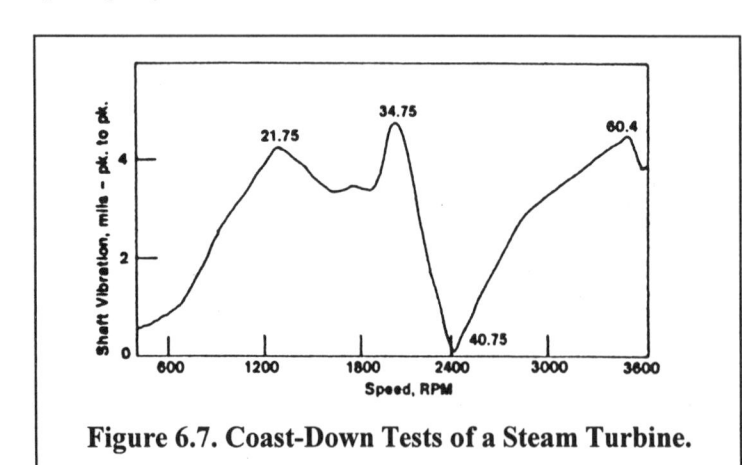

Figure 6.7. Coast-Down Tests of a Steam Turbine.

Run the machine at 10% to 15% over speed if possible, then cut the power and allow the machine to coast down from normal operation as data are being recorded. If either transducers, trigger, or both must be mounted, record the start-up. Run the machine until it is thermally stable before cutting power for the coast-down data. Process the data and identify the critical speeds. Depending on the plot there will be peaks in the spectrum from an FFT analyzer (Figure 6.7), peaks from the tracking filter in a Bodé plot (Figure 6.8), or loops from the tracking filter in a polar plot (Figure 6.9).

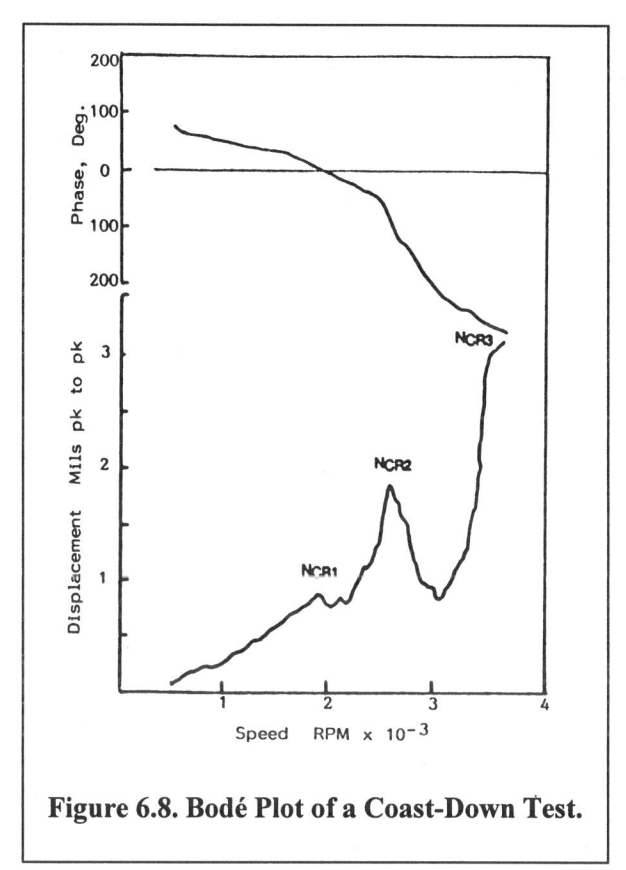

It can be seen that the natural frequency at operating speed is not necessarily the natural frequency measured during the start-up or coast-down test. An interference diagram (see Figure 6.4) is helpful in visualizing the natural frequencies of a machine at speeds other than critical speeds.

Figure 6.8. Bodé Plot of a Coast-Down Test.

Using the FFT Analyzer/Data Collector. The peak hold feature of an analyzer/data collector can be used to provide data on critical speeds. However, the frequency range selected must be sufficiently high to track the coast down. The peak hold feature holds and displays the peak values of all data after each spectrum is computed. The acquisition time of the block of data analyzed is dependent on the frequency span selected. The lower the frequency span, the longer the acquisition time.

The equation for acquisition time T_s is

$$T_s = N/F_{max}$$

N is the number of lines. F_{max} is the frequency span of the analyzer. The acquisition time can be decreased by reducing the lines of resolution. In addition, use of overlap processing also reduces data acquisition time. With overlap processing, a percentage of data from the previous sample is used in calculating the present spectrum. In machine coast-down tests, the process of computing the FFT is always faster than the data acquisition.

If a 400-line analyzer is set on a frequency span of 100 Hz (6,000 CPM), four seconds (400/100) are needed to acquire a sample. Because many samples are necessary to plot a smooth curve during a start-up or coast-down test without data drop out, the frequency span of the analyzer must be evaluated carefully prior to data collection. Resolution is lost with too wide a fre-

6.9

quency span. Too narrow a span may prevent the observation of the critical speed because the data acquisition time is excessive.

Consider, for example, a 12 second start-up of a two-pole motor that operates at 3,600 RPM. If the analyzer is set on 6,000 CPM (100 Hz) and four seconds are required for each sample, only three data points will be obtained — hardly sufficient for a curve. Increasing the frequency span to 400 Hz (24,000 CPM) and decreasing the number of lines to 100 means that a sample is taken every 0.25 second (100/400 = 0.25). A 12-second span/0.25 second/sample is equal to 48 samples, or one sample every 75 RPM. However, the resolution is reduced to increments of 24,000 CPM/100 lines, or 240 RPM, which yields 3,600/240 RPM, or 15 data points over the frequency range between 0 and 3,600 RPM. Eighty percent overlap processing — i.e., 20% data retained — at 200 lines with a 200 Hz (12,000 CPM) span provides a data acquisition time of 0.2 second (200 lines per 200 Hz x 0.2). The resulting 60 points at about 60 RPM intervals with a resolution of 60 CPM may provide an adequate plot. It must be recognized that the small amount of new data contained in each point processed may compromise the analytic results. Under these conditions a tracking filter will provide better results.

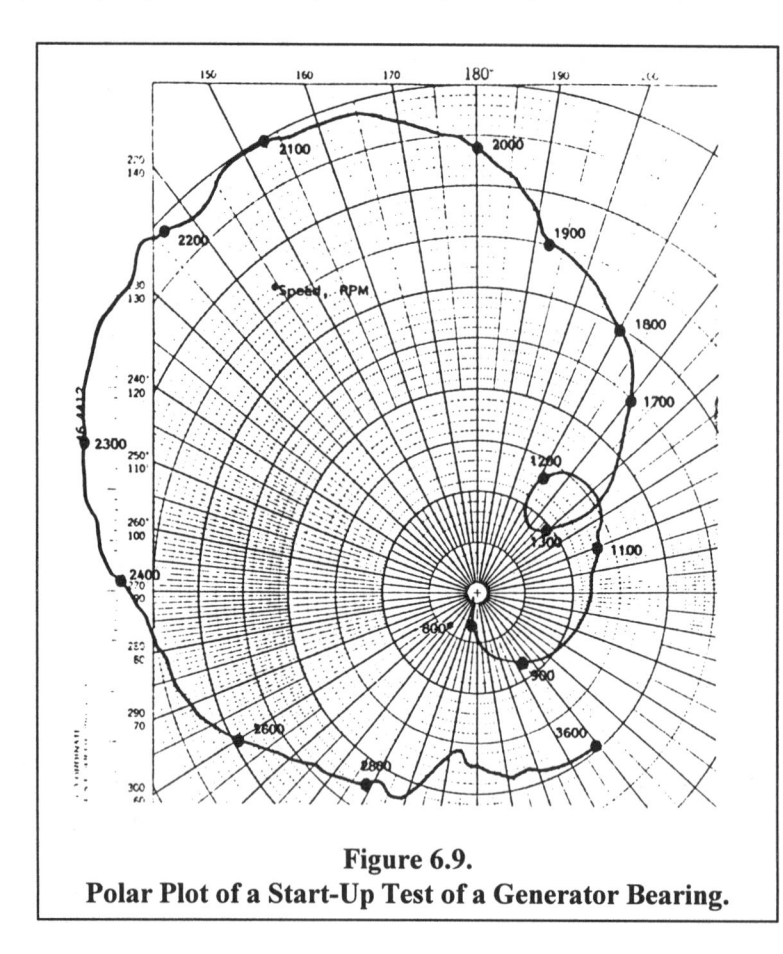

**Figure 6.9.
Polar Plot of a Start-Up Test of a Generator Bearing.**

Figure 6.7 shows coast-down data of a steam turbine analyzed by an FFT. In this example the coast down and start-up required so much time that data acquisition time is not an issue. The dip in this spectrum at 2,400 RPM means that the rotor is bowed.

Using polar plots. Figure 6.9 is a polar plot produced by a synchronous tracking filter from a start-up test of a turbine generator. The plot shows the amplitude and phase of vibration at various speeds. The tracking filter plots the real (amplitude times cosine of phase angle) and the imaginary (amplitude times sine of the phase angle) at the various speeds. The small loop in Figure 6.9 identifies the first critical speed of the generator (1,000 RPM). The large loop is the second critical speed, 2,250 RPM.

In summary, a tracking filter is best for rapid run-up/coast-down tests. Vibration is indicated in the filtered frequency band, which is governed by a reference signal generated by a proximity probe/notch or optical pickup/reflective tape. Peak vibration levels and phase changes indicate critical speeds.

The single-channel analyzer/data collector may be used for impact tests in either time or frequency domains. Triggering can be free or from a hammer source. Vibration peaks indicate resonances. During impact tests a uniform (none) window should be used on the analyzer. Some analyzers have special windows for impact tests.

Fault, Condition, and Balance Tests

Fault analysis and condition evaluation testing are covered in *Chapter IV* and *Chapter V* respectively. Balancing is reviewed in *Chapter VIII*.

Specifications

The purpose of preparing a specification for new or repaired equipment is to procure quality equipment and services, to avoid misunderstandings, to resolve differences of opinion prior to purchase, and to establish a methodology for testing the equipment without controversy. The idea is that everyone participating in the procurement process should understand and agree on the rules of the evaluation.

The acceptable vibration levels specified should be realistic for the type and service of the machine being procured. An overall vibration level of 0.05 IPS_{rms} would not be specified for reciprocating equipment unless it had special isolation mounting.

It is best to use existing API [6.3] or ISO [6.1] standards as guidelines for preparing a specification. The measure specified should be distinct and narrowly defined. Vibration velocity should be specified in distinct IPS units. Is overall peak, derived peak (1.414 x overall rms), component peak, or overall rms with a specified frequency span to be measured? Levels could vary as much as three or four to one, depending on the method used for processing. Without a doubt, instrument-oriented misunderstandings can occur unless the method of processing the signal is distinctly described.

Environment and Mounting

Mounting resonance is often a cause of excessive vibration, especially on vertical pumps. When the natural frequency of a mechanical system is equal to or close to the operating speed of the pump or two times the operating speed, resonance occurs and thus amplification of vibration. Resonance amplifies vibration as a result of mass unbalance and hydraulic forces that might occur normally. Careful design and testing by the pump manufacturer results in natural frequencies of the shaft/impeller-oriented pump and the pump drive line that are out of the range of operating speeds. Unfortunately, the manufacturer generally has no control over mounting and piping arrangements and cannot be responsible for natural frequencies of the entire system. It is the responsibility of the customer to assure that the architect-engineer understands resonance and its consequences. Sufficient bracing and support should be used in piping to assure a natural frequency higher than pump specifications.

Presentation of Data

The way data are presented will determine whether or not good data, properly processed, will be valuable for fault analysis, condition evaluation, and baseline testing. Data from acceptance testing is generally in a simple form that involves simple overall levels. Spectral data that are properly presented provide the resolution and dynamic range sufficient to discern important frequencies and amplitudes. The time waveform should be presented so that the data can be related to the physical characteristics of the machine.

Detailed waveform variations should be observable; otherwise, long-term amplitude trends may be required that lead to multiple processing in the FFT. The span of the time waveform is equal to the data acquisition time from the analyzer; that is, number of lines divided by the frequency span. Orbits must not be filtered at operating speed if diagnostics are required. High-frequency filtering may be required to remove noise. However, this process can introduce amplitude and phase errors. The phase of operating speed vibration to a spot on the shaft is a valuable piece of information for analysis. Figure 6.10 is a spectrum and time waveform for an electric motor. These data show long-term trends in the time waveform. However, the resolution in the spectrum is insufficient to resolve two times the operating speed (RPM) and two times line frequency (120 Hz) or the vibration at operating speed. The variation of the amplitude in the time waveform leads the analyst to realize that the components in the bins at 60 Hz and 120 Hz are either varying in amplitude or multiple frequency components exist. Figure 6.11 and Figure 6.12 show sidebands around each of the components.

Figure 6.10. Spectrum and Time Waveform for an Electric Motor.

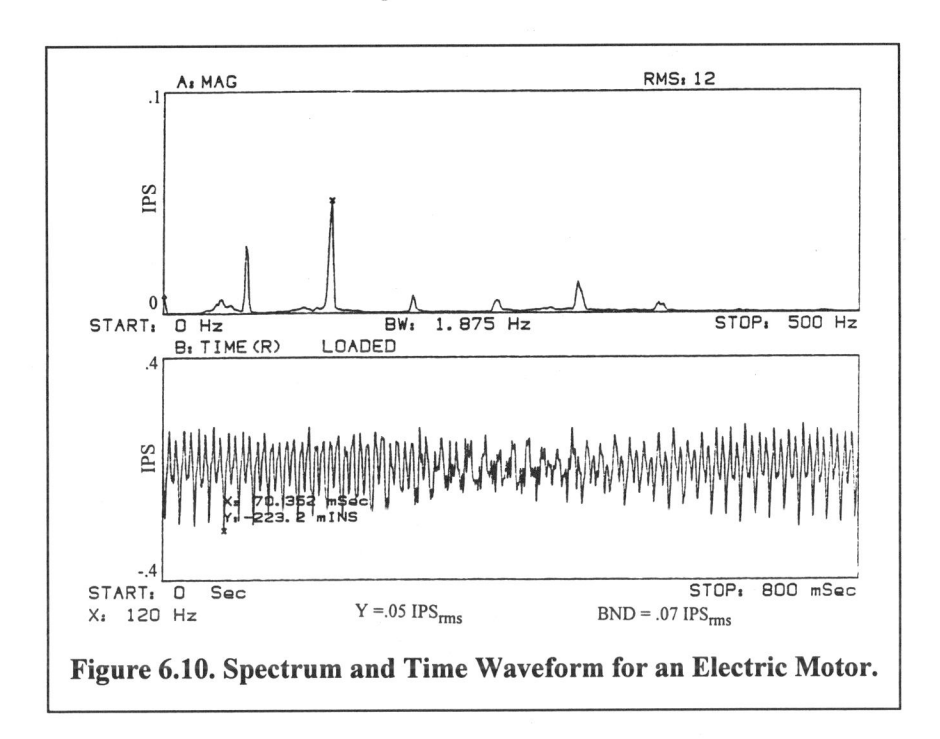

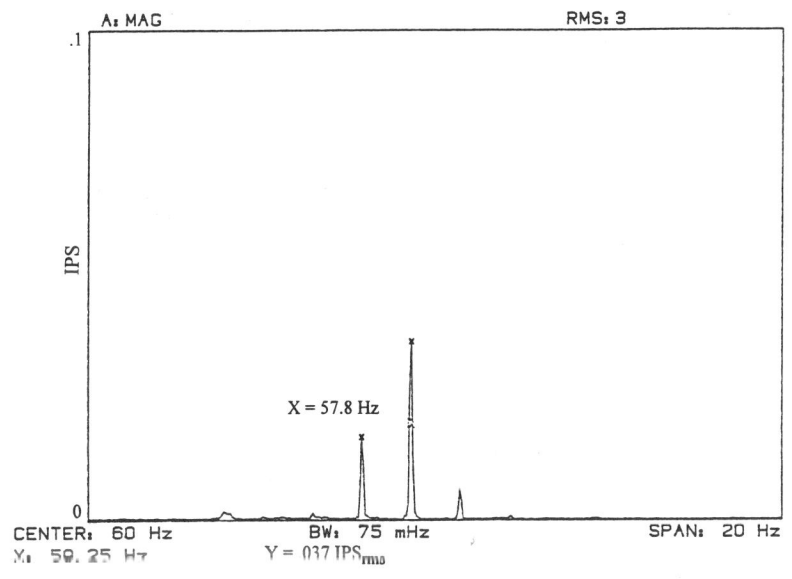

Figure 6.11. Sidebands Around the Component in the 60 Hz Bin.

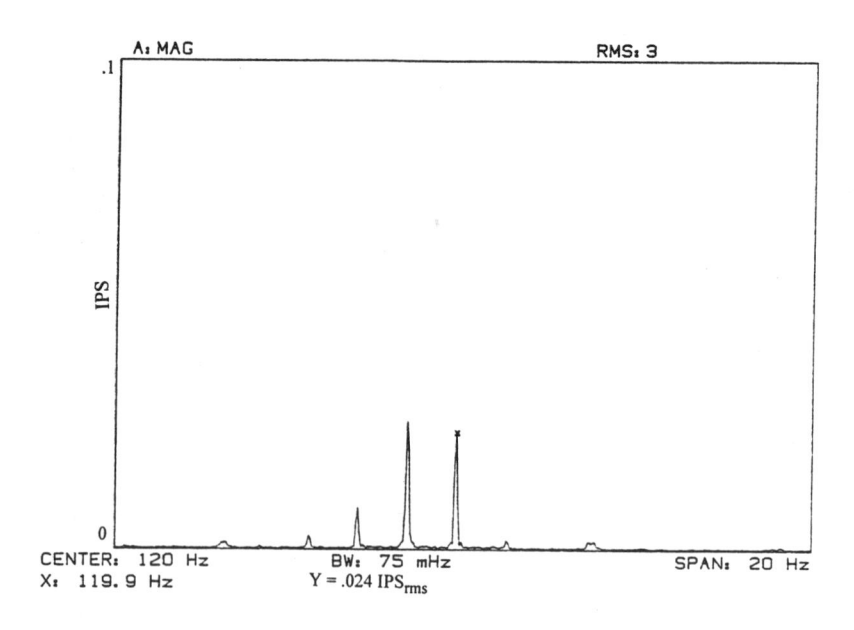

Figure 6.12. Sidebands Around the Component in the 120 Hz Bin.

Figure 6.13 contains data from a fan motor in which the time waveform and spectrum were processed separately to obtain resolution and shape in both displays. The data in Figure 6.14 show the spectrum from a generator displayed in terms of orders. This display is advantageous for a variable-speed machine because the spectral components do not smear.

In conclusion, data should be presented so that frequencies can be resolved and amplitudes accurately defined. In addition, it must be possible to relate the shape of the time waveform to the physics of the machine.

Reports

Reports should be written for each activity — baseline listing, acceptance testing, or in-depth analysis, which would include operational tests, resonance and critical speed tests, and environmental tests. Reports should be well organized, brief but complete, and contain the following sections: Executive Summary, Introduction, Technical Discussion, Conclusions and Recommendations, and Appendix (Technical Data).

The Executive Summary should contain a description of the equipment being tested, the symptoms of the problem, major findings, and conclusions and recommendations. It should be brief but descriptive, so that management can base decisions on the situation without reading the entire report. The Introduction should describe the equipment being tested, purpose of the test, approach to the test, and test equipment and techniques used.

Technical details supporting the conclusions and recommendations are presented in the Technical Discussion, as are selected supporting data. All data should be included in the Appendix, but only such specific information as a description of measurement points, last measurement reports, trend plots, and spectral data on exceptions and alarms should be included in a routine periodic monitoring report. An in-depth analysis report should include selected time waveforms, spectra, orbits, and test data from impact or coast-down tests. Baseline test reports should provide a complete picture of the condition of the equipment or faults present. Baseline tests contain the normal vibration levels to the best of the capability of the analyst and suggest values for setting alarms.

The acceptance test report should be linked to the specification or the wishes of the operator. It may be necessary to carry out a complete analysis of the machine during acceptance testing. Reports generated after balancing should show vibration levels and the trial weights applied as the balance progresses. After the unit is balanced, final readings should be recorded as well as the balance sensitivity and phase lag values.

Conclusions and recommendations are necessary for all major findings found in the survey or analysis. They should be brief but inclusive.

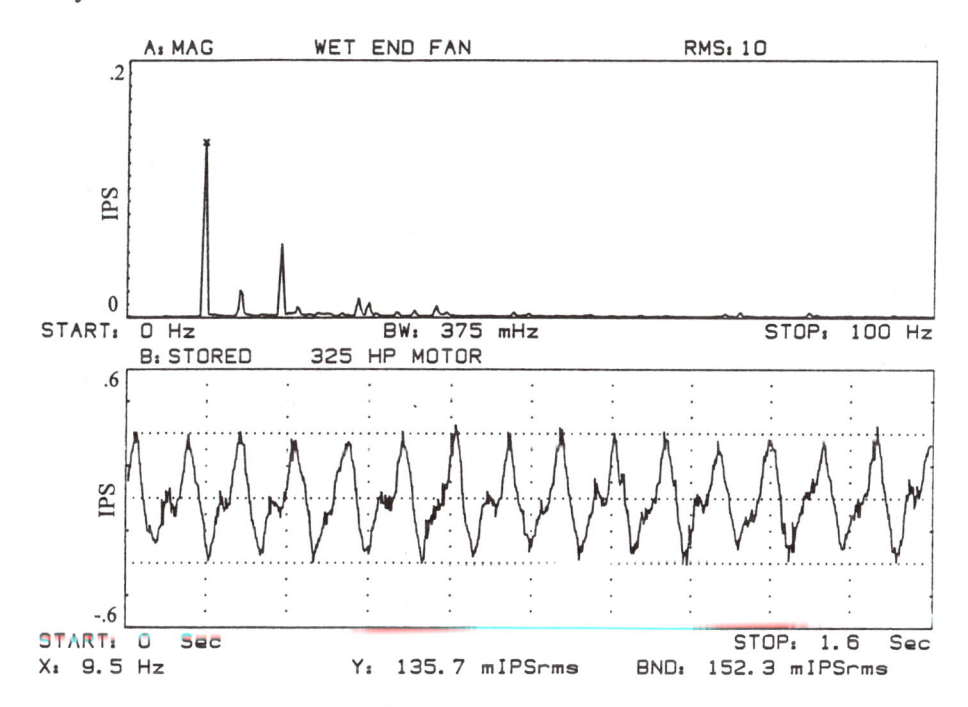

Figure 6.13. Data from a Fan Motor.

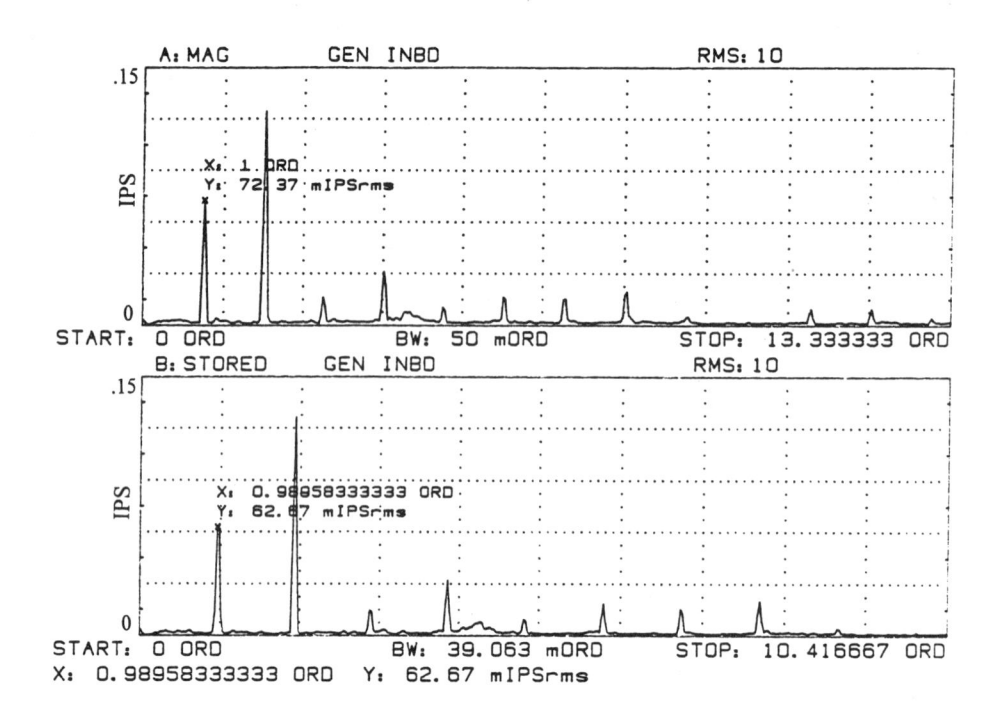

Figure 6.14. Generator Spectrum in Terms of Orders.

Summary of Machine Testing

- A test plan should be generated prior to data acquisition on a machine — acceptance tests, baseline tests, fault analysis, condition evaluation, design, and balancing.
- The test plan should contain a description of the machine, the tests to be performed, the data to be acquired, loads, speeds, machine configurations, and process conditions.
- The data acquisition plan should provide details about sensors including location, measures, and process conditions.
- If data are processed on site, the analyzer setups must be provided, including frequency spans, lines of resolution, range, windows, and time spans. Sometimes multiple data acquisitions are required to obtain adequate range and resolution.
- A site inspection should provide details about external vibrations and machine mounting.
- Acceptance tests are to be listed in detail in the purchase specification of a new or repaired machine. Included are procedures, measurement locations, process conditions, measures and the way they are processed, and acceptable vibration levels.
- Baseline tests are conducted to establish normal operating levels of vibration when the machine is in good operating condition.

- Specifications should be used to assure the procurement of quality equipment.
- Be realistic about acceptance levels and locating critical speeds.
- Good mounting environments and procedures will assure that equipment is operating properly.
- Presentation and reporting of data provide quality analysis of quality data.

References

6.1. ISO 2372, 1974, "Mechanical Vibrations of Machines with Operating Speeds from 10 to 200 RPS — Basis for Specifying Evaluation Standards," International Standards Organization, Geneva, Switzerland (1974).*

6.2. Maedel, P. H. Jr., "Vibration Standards and Test Codes," *Shock and Vibration Handbook*, 4th Edition, C.M. Harris, Editor, McGraw-Hill, NY (1996).

6.3. American Petroleum Institute Procurement Standards, API, Washington, D.C.

* ISO Standards can be obtained from the Director of Publications, American National Standards Institute, NY, NY 10005-3993.

CHAPTER VII
PERIODIC MONITORING

Cost-effective monitoring means higher profits, recognition, and a better quality of life.

Since it was initiated in the 1970s, periodic monitoring of machine vibration has become the principal component of predictive maintenance programs in many industries. The advent of the electronic data collector has made cost effective the routine collection, trending, and analysis of data. One individual can effectively monitor the thousands of data points that relate the condition of many machines. On the other hand, continuous monitoring provides protection and the capability to evaluate critical equipment [7.1]. Oil analysis, thermography, and electric current monitoring are used in conjunction with vibration analysis in noninvasive predictive maintenance programs. This chapter on periodic monitoring includes information about program development, including listing and categorization of machines, route determination, measurement points, measures, baseline data, frequency of data collection, trending, alarms, recommended maintenance actions, and reports.

Machines are selected for monitoring and their monitoring priorities are established before detailed plans are made. Baseline data are used to define the normal operating conditions for a machine and to establish the data needed for effective monitoring. The goal of any monitoring program is to select measurements that provide the greatest sensitivity to any change in machine condition but are not complex and do not require extensive data processing. A procedure should be chosen for each machine when a program is begun and then modified as new information is obtained. The default monitoring setup is often overall vibration levels at two points — radial and axial — on each bearing on a quarterly basis. However, data are often taken monthly from the horizontal, vertical, and axial positions in a new program. In most cases, after data are taken for some time, the number of data points may be decreased but more than one type of data may have to be taken more often. In addition, the sophistication of the measure may increase from peak or rms overall levels to band-filtered processing, spectral evaluation, or demodulation. Such decisions are based on the experience gained as the program evolves. Cost justification and performance are very important; otherwise, management may lose interest and underfund or eliminate the program.

Listing and Categorization

A list of the machines in the plant is the first step in a monitoring program. The machines should be categorized according to a hierarchy based on the criticality of the machine to plant operation. Table 7.1 is a method of categorization developed by the petrochemical industry. The machines are ranked in four grades — A, critical; B, critical or failure-prone; C, critical spared; and D, noncritical. The periodic monitoring program should initially focus on A and B.

Depending on the resources available, machines in class C can be included in the program at a later time. Each plant or mill has its class A machines — turbines and compressors for the petrochemical industry, turbine generators for the power industry, and paper machines for the pulp and paper industry. Unexpected failure of a single component can result in the loss of millions of dollars due to production downtime.

Table 7.1. Machinery Classification for Monitoring.

Machinery Classification	Result of Failure
A. Critical	unexpected shutdown or failure will cause significant production loss
B. Critical or Failure-Prone	unexpected shutdown or failure reduces but does not interrupt production
C. Critical Spared	light-duty service-causes inconvenience in operation but no interruption of production; repair costs justify some level of monitoring
D. Noncritical	production will not be affected by loss; repair cost does not justify monitoring

Machinery Knowledge

Knowledge of the characteristics of machinery is essential to conducting efficient vibration analyses. The more information available about the machine design, construction, supports, operational responses, and defect responses, the easier will be the diagnosis of defects and malfunctions. All service equipment should be cataloged and the following data listed.

- broad characteristics such as rotational frequencies, gear mesh, vane pass, and bearing defect frequencies.
- vibration, temperature gradients, or pressure initiated by an operating component or system.
- vibration responses to process changes.
- characteristics identified with the specific machine type.
- known natural frequencies and mode shapes.
- sensitivity to instability from wear or changes in operating conditions.
- sensitivity to vibration from mass unbalance, misalignment, distortion, and other malfunction/defect excitations.

Certain responses (Table 7.2), including vibration, temperature, and pressure can be related to components of the system; e.g., bearings (rolling element, hydrodynamic journal, and thrust); balance drum of a centrifugal compressor; blading; and gear mesh and gear teeth. Frequencies and their components, including sidebands, indicate the existence of a fault and its source. Shaft speeds usually modulate the center frequencies of bearings and blades. The amplitudes of sidebands and center frequencies provide information about severity. Temperature and pressure readings may be present as DC components.

Table 7.2. Component Sources of Machine Excitation and Response.

Component	Frequency
antifriction bearings	ball pass frequency, outer race ball pass frequency, inner race fundamental train frequency rotating unit frequency ball spin frequency
hydrodynamic journal bearings	fractional frequency whirl frequencies
gears	rotating unit frequency gear-mesh frequencies and harmonics harmonics of gear-mesh frequencies assemblage frequencies system natural frequencies (gear-tooth defects)
blade wheels and impellers	rotating unit frequencies vane and blading frequencies harmonics of vane and blading frequencies
rotors	trapped fluid rotational frequency directional natural frequencies higher harmonics
couplings and universal joints	orders of rotating frequency
reciprocating mechanisms	rotating frequency and its orders

Characteristics associated with specific machines are listed in Table 7.3. The natural frequencies and mode shapes of equipment can provide valuable information during the diagnosis of a machine problem. This information is valuable in predicting wear, product buildup, corrosion, looseness, thermal changes, and other malfunctions. Natural frequencies and mode shapes are commonly calculated by the vendor or user from mass-elastic data.

Table 7.3. Characteristics of Machine Excitation and Response.

Machine Type	Characteristics
Centrifugal Machines — Impeller Types	
centrifugal pumps	stiff bearings vane-passing frequencies and their multiples
centrifugal compressors	large casing-to-rotor weight ratio sleeve or tilt pad bearings vane-passing frequencies from impellers
fans	vane-passing frequencies system aerodynamics pedestal characteristics often important
Bladed Machines	
axial-flow compressors	blade-passing frequencies and their multiples
steam turbines — mechanical drives	blading frequencies 5,000 RPM-12,000 RPM range for critical speeds rubs and mass unbalance
steam turbines for power generation	blading frequencies low-speed — large casings high pressures mass unbalance
gas turbines	blading and gearing frequencies subject to instabilities and rubs
Power Transmission Equipment	
gearboxes	gear-mesh frequency and higher harmonics casing resonances gear-tooth excitation of natural frequencies pitch-line runout torsional responses
fluid drives	slip-frequency excitation bearing whirl
Motors/Generators	slip frequency modulation pole-induced structural vibration thermal-induced excitation high synchronous-motor excitation at start-up stator shorts
Reciprocating Machines	
engines	casing distortion bearing-induced foundation vibration high torsional excitation by inertia and pressure
pumps and compressors	high torsional excitation by inertia and pressure
Small Equipment	antifriction bearing failures looseness belts and gear-drive problems

Details of each machine should be available before the database is used. Figure 7.1 is an example of a data sheet. The location and name of the equipment are needed, as are the author and date on which the information was acquired. Space is provided for a generic description of the machine; e.g., locally-derived terminology that accurately describes the machine and its configuration. Plant asset numbers, if available, can help avoid confusion with regard to missing or duplicated machines. There is space on the sheet for a sketch of the installation for reference and for general background and specific mechanical information.

The mechanical component list is critical in structuring the database and for follow-up analysis. Minimum requirements for electric motors include horsepower, operating speed, and specific information about the bearings; this information can usually be obtained from the name plate or the distributor. The motor type (AC, DC, synchronous, induction, inverter driven) is used to select measurement ranges and to provide basic analytical information.

Specific information about reduction gear units is usually more difficult to acquire. The name plate information (manufacturing model and serial number) provides documentation. Additional information, which includes internal configurations and identification of components of bearings and gear teeth, can usually be obtained from the manufacturer through a sales representative or distributor. It is always advisable to request a general arrangement drawing that shows the internal orientation of specific components (see Figure 7.2).

Other machines that may be part of a predictive maintenance program are pumps, machine tools, process mixers, grinders, crushers, turbines, mills, and suction rolls. Mechanical component data include general arrangement sketches, bearing identification, operating speed or range of speeds of the machine, and type of driver. Any idiosyncrasies relative to the machine — mountings, obvious weaknesses, and process conditions that can affect operation and diagnostics — should be noted.

Route Selection and Definition

The route selected for collecting data can be based on plant layout, machine train (process stream), machine type, or type of data required. Plant layout and machine train routes are the most commonly used. Routes based on plant layout follow the floor plan and progression from one machine to another. Basing a route on a machine train means that data are taken on all machines in the production or processing of a product line, regardless of physical placement. For

SURVEY REQUEST

PLANT: _____ BY: _____ DATE: _____

AREA: _____ EQUIPMENT: _____

GENERAL DESCRIPTION: ASSET NUMBER: _____

_____
_____
_____
_____

MECHANICAL COMPONENT
INFORMATION:

ROLL DIAMETER: _____ AVG. MACH. SPEED: _____

BEARINGS (MOTOR, ROLL) MOTOR NAMEPLATE DATA:

INBOARD _____ H.P. _____ S.F. _____

OUTBOARD _____ RPM _____ FRAME _____

REDUCER: TEETH   BEARINGS

TYPE: _____ 1ST. RED: _____ _____

RATIO: _____ 2ND RED: _____ _____

MFG.: _____ 3RD RED: _____ _____

MODEL #: _____

SERIAL #: _____

Figure 7.1. Example of a Survey Request Form.
Courtesy of Mechanical Consultants, Inc.

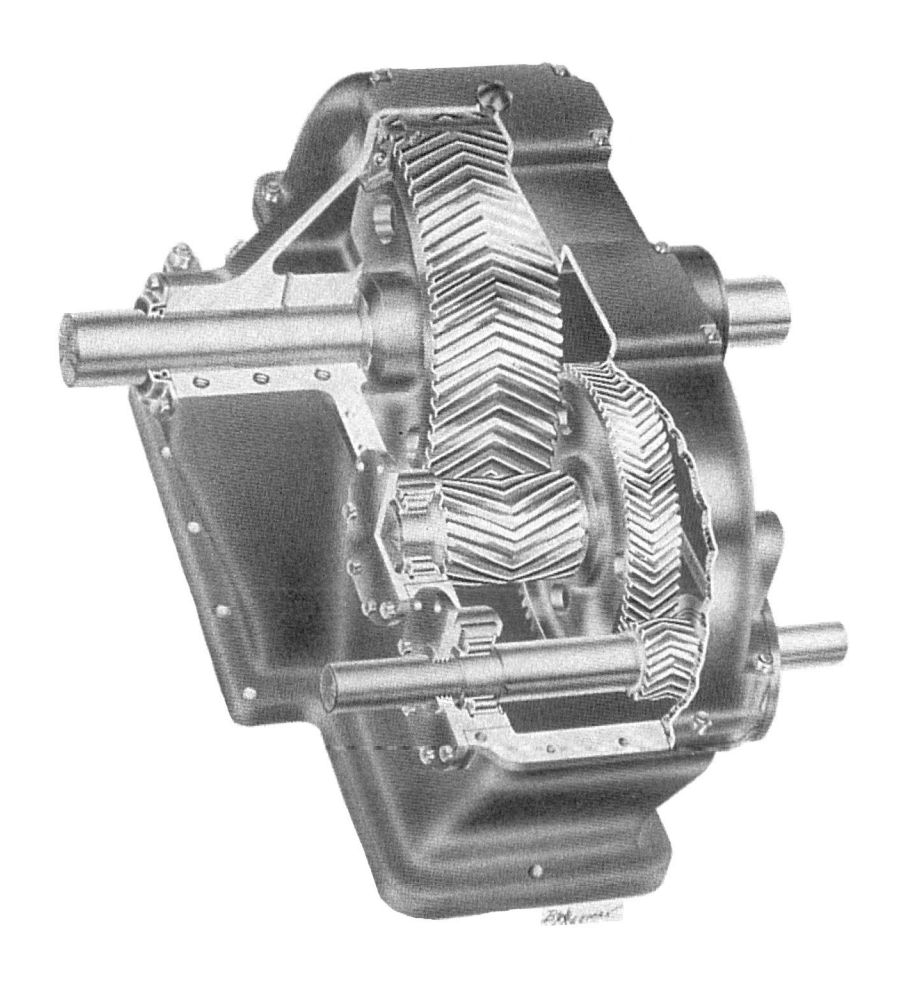

Figure 7.2. Double Reduction Gearbox.
Courtesy of Lufkin Industries, Inc.

example, the route of a turbine generator includes the boiler feed pumps, lube oil pumps, ID fans, and circulating water pumps. If the route is based on machine type, all machines of the same type, for instance, electric motors of a given size or bearing type, would be included; similarly, vertical multistage pumps would be measured as a group. Data routes require that all points have similar processing — spectral, overall, band, or HFD — or similar components; permanently-mounted sensors in a control room are in this category. Any route should be set up so that it is natural and easy to follow from machine to machine; e.g. radial-axial measurements on each bearing. Various aspects of plant layout and access to the machinery are part of the route selection process. Routes should be tailored to meet the needs of the plant, the equipment, and the operator. The initial route when a program is begun should be a small number of machines, fewer than five.

Measures and Measurement Points

Measurement points are identified after the route has been selected. Figure 7.3 shows a method for identifying the machine unit by number and type, measurement locations, transducer directions, and the measure. This route is for casing measurement on the small turbine gearbox-driven generator shown schematically in Figure 6.1. Each line in Figure 7.3 describes one measurement. The data in Figure 7.3, last measurement report, are the unit ID, location and position of the transducer, measure unit, date of measurement, previous amplitude of vibration, last amplitude of vibration, percentage of change, and alarm status. There was a separate route for the permanent monitoring system.

Figure 7.3. Example of a Route for Small Turbine Gearbox-Driven Generator.

```
SKF 17-NOV-94                    *** MCI DATABASE ***
MMMMMMMMMMMMMMMMMMMMMMMMMMMMMMMMMMMMMMMMMMMMMMMMMMMMMMMMMMMMMMMMMMMMMMMM

                              Last Measurement

Id                      Units       Date      Prev Val   Last Val   %Chg   Alrm Sta
DDDDDDDDDDDDDDDDDDDD    DDDDDDDDDD  DDDDDDDDD DDDDDDDDD  DDDDDDDDD  DDDDD  DDDDDDDDD
*** UNIT 2 TUR/GEN STD
2 TUR IN   HOR          IPS         17-NOV-94 0.43606    0.4828     11     A2
2 TUR IN   VER          IPS         17-NOV-94 0.1559     0.06862    -56    ---
2 TUR IN   AX           IPS         17-NOV-94 0.24359    0.2933     20     A1
2 TUR OUT  HOR          IPS         17-NOV-94 0.20413    0.211      3      A1
2 TUR OUT  VER          IPS         17-NOV-94 0.27773    0.1691     -39    ---
2 TUR OUT  AX           IPS         17-NOV-94 0.24792    0.1529     -38    ---
2 RED IN   HOR HFD      G HFD       17-NOV-94 0.61296    1.303      112           P
2 RED IN   HOR ACC      Gs          17-NOV-94 1.06701    1.383      30     ---
2 RED IN   HOR VEL      IPS         17-NOV-94 0.05997    0.06126    2      ---
2 RED IN   VER HFD      G HFD       17-NOV-94 0.35645    1.832      414    A1    P
2 RED IN   VER ACC      Gs          17-NOV-94 1.10673    3.908      253    A1    P
2 RED IN   VER VEL      IPS         17-NOV-94 0.08385    0.08152    -3     ---
2 RED IN   AX  HFD      G HFD       17-NOV-94 0.98513    0.871      -12    ---
2 RED IN   AX  ACC      Gs          17-NOV-94 1.31834    1.113      -16    ---
2 RED IN   AX  VEL      IPS         17-NOV-94 0.09144    0.07465    -18    ---
2 RED OUT  HOR HFD      G HFD       17-NOV-94 0.46739    0.929      99           P
2 RED OUT  HOR ACC      Gs          17-NOV-94 0.8527     1.575      85           P
2 RED OUT  HOR VEL      IPS         17-NOV-94 0.07275    0.06803    -6     ---
2 RED OUT  VER HFD      G HFD       17-NOV-94 0.67461    0.502      -26    ---
2 RED OUT  VER ACC      Gs          17-NOV-94 1.19957    2.112      76     A1    P
2 RED OUT  VER VEL      IPS         17-NOV-94 0.02568    0.06956    171          P
2 RED OUT  AX  HFD      G HFD       17-NOV-94 0.96285    1.598      66     A1    P
2 RED OUT  AX  ACC      Gs          17-NOV-94 2.94848    2.722      -8     A1
2 RED OUT  AX  VEL      IPS         17-NOV-94 0.09342    0.06782    -27    ---
2 GEN IN   HOR          IPS         17-NOV-94 0.0236     0.01579    -33    ---
2 GEN IN   VER          IPS         17-NOV-94 0.03745    0.01634    -56    ---
2 GEN IN   AXIAL        IPS         17-NOV-94 0.02009    0.0472     135    ---
2 GEN OUT  HOR          IPS         17-NOV-94 0.0188     0.02154    15     ---
2 GEN OUT  VER          IPS         17-NOV-94 0.01567    0.01533    -2     ---
2 GEN OUT  AXIAL        IPS         17-NOV-94 0.01296    0.02352    81     ---

30 points printed.
```

The figure shows overall seismic velocity measures on the turbine, generator, and gearbox bearing housings; and peak acceleration and HFD measures on the gearbox bearing housings. Band or component values might have been needed if rolling element bearings were present. Measurement points [7.2] are always selected as close to the bearing as possible.

In a new program, data would be taken in the horizontal, vertical, and axial directions. As experience is gained, the number of measurement points is usually reduced when redundant data are being collected. Data are collected to assess machine condition, not to fill the requirements of a rigid program. Redundant data are not cost effective.

Measures are intended to be sensitive to machine condition and are selected on the basis of machine speed, component frequencies, and process characteristics (*see Chapter II*). Frequency spans for monitoring can be adjusted according to the results of the baseline data.

A minimum of two radial readings and an axial reading should be taken on electric motors, depending on the size of the motor and operating conditions. The readings should be in the plane with the greatest flexibility; that is, the position most likely to respond to forces generated by the machine. The horizontal plane on the drive end and idle end would be used for a standard motor mounted to a conventional baseplate/foundation. Depending on motor speed, filtered high-frequency responses can also be useful in determining such operating parameters as lubrication, bearing-surface condition, and other mechanisms that generate higher-than-normal frequencies. Six measurements over three points may be necessary.

Typical setups for a high-frequency measurement and a velocity measurement on an electric motor are shown in Figure 7.4 and Figure 7.5. The setups allow for two types of measurement at one point on the machine with the same transducer. The optimum point configuration provides the following.

- measurement that is a response to machine condition.
- alarm levels that announce threshold crossing and initiate analytical data collection.
- time waveform and spectral data with 400 lines of resolution upon alarm.
- frequency ranges that allow for analysis of operating speed orders (RPS) and likely bearing-defect frequencies on one spectrum.

A measurement point is assigned for each end of a motor, each rotational component of a reducer/increaser, and each end of a roll with a bearing at each end. The orientation of the transducer may be radial, axial, or both, depending on such factors as machine type, bearing type, service, and application. The points are organized under the respective machine, area, and plant into a hierarchy of the database. Figure 7.6 is a simplified hierarchical outline containing a

```
                         ** MCI DATA BASE **
MMMMMMMMMMMMMMMMMMMMMMMMMMMMMMMMMMMMMMMMMMMMMMMMMMMMMMMMMMMMMMMMMMMMMM
IMMMMMMMMMMMMMMMMMMMMMMMMMMMMMMM HIERARCHY MMMMMMMMMMMMMMMMMMMMMMMMMMM;
I                                          NEWS MILL                  I
I P/M DRIVE SYSTEM                         D.C. MOTOR/REDUCTION UNITS I
I TOP WIRE DRIVE                                                      I
LMMMMMMMMMMMMMMMMMMMMMMMMMMMMMMMM DESCRIPTION MMMMMMMMMMMMMMMMMMMMMMMM9
I Id:TOP WIRE MTR IER HFD              Last modified on: 13-DEC-88 11:56:04 I
I Description:TOP WIRE DRIVE                                          I
I Schedule:90 days                     Down Load: Enabled             I
LMMMMMMMMMMMMMMMMMMMMMMMMMMMMMMMM INPUT SETUP MMMMMMMMMMMMMMMMMMMMMMMM9
I Point type: Peak HFD                 Full scale:2         G HFD     I
I Detection:  PEAK       Input mV/EU:100                              I
I                                                                     I
I                                                                     I
I                                                                     I
I                                                                     I
I                                                                     I
I                                                                     I
LMMMMMMMMMMMMMMMMMMMMMMMMMMMMMMMM ALARM SETUP MMMMMMMMMMMMMMMMMMMMMMMM9
I Type: LEVEL            Lower:0.75             Upper:1.5             I
I                                                                     I
MMMMMMMMMMMMMMMMMMMMMMMMMMMMMMMMMMMMMMMMMMMMMMMMMMMMMMMMMMMMMMMMMMMMM<
```

Figure 7.4. Setup for a High-Frequency Measurement on an Electric Motor.

```
                         ** MCI DATA BASE **
MMMMMMMMMMMMMMMMMMMMMMMMMMMMMMMMMMMMMMMMMMMMMMMMMMMMMMMMMMMMMMMMMMMMMM
IMMMMMMMMMMMMMMMMMMMMMMMMMMMMMMM HIERARCHY MMMMMMMMMMMMMMMMMMMMMMMMMMM;
I                                          NEWS MILL                  I
I P/M DRIVE SYSTEM                         D.C. MOTOR/REDUCTION UNITS I
I TOP WIRE DRIVE                                                      I
LMMMMMMMMMMMMMMMMMMMMMMMMMMMMMMMM DESCRIPTION MMMMMMMMMMMMMMMMMMMMMMMM9
I Id:TOP WIRE MTR IER                  Last modified on: 10-FEB-89 15:40:00 I
I Description:TOP WIRE DRIVE                                          I
I Schedule:90 days                     Down Load: Enabled             I
LMMMMMMMMMMMMMMMMMMMMMMMMMMMMMMMM INPUT SETUP MMMMMMMMMMMMMMMMMMMMMMMM9
I Point type: Velocity (Acc to Vel)    Full scale:0.2        IPS      I
I Detection:  RMS        Input mV/EU:100      Low Freq. Limit:2       I
I                                                                     I
I                                                                     I
LMMMMMMMMMMMMMMMMMMMMMMMMMMMMMMMM FFT SETUP MMMMMMMMMMMMMMMMMMMMMMMMMM9
I RPM:1200     Lines: 400  Freq type: FIXED SPAN   Frequency:1000  Hz I
I Averages:8             Window: HANNING    Auto Capture: ON OVERALL ALARM I
I Probe dir: VER. RADIAL                 Storage depth:55             I
LMMMMMMMMMMMMMMMMMMMMMMMMMMMMMMMM ALARM SETUP MMMMMMMMMMMMMMMMMMMMMMMM9
I Type: LEVEL            Lower:0.07             Upper:0.2             I
I                                                                     I
MMMMMMMMMMMMMMMMMMMMMMMMMMMMMMMMMMMMMMMMMMMMMMMMMMMMMMMMMMMMMMMMMMMMM<
```

Figure 7.5. Setup for a Velocity Measurement on an Electric Motor.

```
                        ** MCI DATA BASE **
HHHHHHHHHHHHHHHHHHHHHHHHHHHHHHHHHHHHHHHHHHHHHHHHHHHHHHHHHHHHHHHHHHHHHHHH!
                          HIERARCHY OUTLINE

MCI SYSTEMS DEMO         .................  Demo database
   PAPER CORP               .................  NEWS MILL
      RANDOM DATA FILE  ................  MISCELLANEOUS DATA
      P/M DRIVE SYSTEM       ...............  D.C. MOTOR/REDUCTION UNITS
         TOP WIRE DRIVE        .............
            TOP WIRE MTR IER HFD .........  TOP WIRE DRIVE
            TOP WIRE MTR IER       .........- TOP WIRE DRIVE
            TOP WIRE MTR DER HFD .........  TOP WIRE DRIVE
            TOP WIRE MTR DER       .........  TOP WIRE DRIVE
            TOP WIRE MTR DEX       .........  TOP WIRE DRIVE
            TOP WIRE MTR DEX HFD .........  TOP WIRE DRIVE
            TOP WIRE RED IN RAD  .........  TOP WIRE DRIVE
            TOP WIRE RED IN AX   .........  TOP WIRE DRIVE
            T/WIRE RED IN AX HFD .........  TOP WIRE DRIVE
            T/WIRE RED OUT RAD   .........  TOP WIRE DRIVE
            T/WIRE RED OUT AX    .........  TOP WIRE DRIVE
            T/WIRE RED OUT AX HF .........  TOP WIRE DRIVE
```

Figure 7.6. Simplified Outline of a Hierarchy.

description of a plant, the highest category; a subsystem of machines, the second category; a single machine from the group at the next level; and, finally, individual measurement points that are used to evaluate the machine.

Baseline Data

Baseline conditions are necessary to evaluate the condition of a machine because machines of the same design operate at different normal vibration levels. Variations in installation conditions, including alignment, piping, and foundations, are the cause. Thus, severity levels are useful only as guidelines for condition evaluation.

Baseline data provide the initial data for selecting a trending type and the trend database, as well as information for setting alarms. Trending of overall measures is typical, but trending in bands may be necessary. A spectrum and time waveform should be recorded at each point. If data are taken from displacement probes, the amplitude and phase at operating speed (with respect to the shaft trigger) should be taken for class A equipment 500 HP and above (see Table 7.1). Transient data are valuable for this type of equipment. Start-up and coast-down data should be taken on these machines. Transient data include a Bodé plot (magnitude of displacement and phase vs machine speed) and a polar plot (amplitude and phase at various speeds). Both plots show the critical speeds of the machine. The data are used for balancing and damping evaluation.

Frequency of Data Collection

The mean time to failure of machine components, criticality of the machine, number of spares, production and repair costs of failure, available personnel, and monitoring costs are factors in determining the frequency of monitoring. No arbitrary span such as a week, month, or several months is possible. The records for a machine should be reviewed to assess failure frequency in the past. Important factors are costs in loss of production and machine replacement as well as personnel costs. If there is no spare, monitor more frequently, perhaps once a month. Monitor less frequently if a machine is performing well.

Perform the program properly. If the schedule does not allow for consistent work, decrease the number of machines monitored. Chronic problems with machines should be solved. Periodic monitoring is an expensive way to compensate for unreliable machines. With the exception of critical machines, quarterly monitoring should be adequate; important factors are reliability and operating speed of the machine. High-speed machines cycle many times in a short time and may require more frequent monitoring. Extrapolated trends for a machine are not dependable unless vibration data during failures have been documented.

Selection of Test Equipment

The selection of test equipment depends on the operation of the program, the number of data points, and the depth of the analysis. If the test equipment was purchased before the program was planned, it may have to be tailored to the plant equipment. Otherwise, the equipment and computer software can be chosen to meet the needs of the program. If there is a question concerning the proper test equipment, select instruments that are versatile and dependable. At a minimum the data collector and computer software should be capable of trending overall and band readings; performing analog and digital integration; providing up to 6,400 lines of resolution with a dynamic range of 72 dB; selecting Hanning or uniform windows; and performing overall HFD measurements and demodulated high-frequency spectra with selectable filters [see Reference 7.1 for additional details].

Screening

Screening is used to assess at relatively low cost when a problem is developing in a specific machine. It allows time for an analysis and to prepare for repairs. Screening techniques vary in sophistication and effectiveness. Effectiveness depends on the device used and the type of machine being monitored. Allowances must be made for changes in operating conditions that affect overall vibration levels. Such changes may be caused by changes in the process or environmental conditions. Any mechanism that can be used to relate changes in vibration due to process changes increases the effectiveness of a program. Trends should based only on changes in machine condition if possible. It is thus good monitoring policy to carry out a thorough vibration analysis before initiating a maintenance action.

Simple methods. Early screening devices included screw drivers, wires, and stethoscopes. They were used to detect faults in rolling element bearings. A screening device such as a true rms meter and velocity transducer is simple but labor intensive. The rms can be calculated by an FFT analyzer with data acquired from a velocity transducer or an accelerometer, depending on the machine. A doubling of vibration level usually indicates that some action is necessary, either a more detailed vibration analysis or initiation of repair.

Other simple instruments — e.g., single-value high-frequency defect meters — use accelerometer response to pulses resulting from a distinct fault in a specific machine component. The response of an accelerometer is filtered to include only the activity around its natural frequency.

Simple instruments are adequate screening devices if nondestructive pulses and noise are not present at the measurement point. For example, the changes in vibration level that occur when faults arise in a well-balanced motor-driven fan with a low vane-pass frequency will be detected because bearing-associated pulses can be sensed. However, a pulse-indicating instrument may not be capable of distinguishing a new fault if the pulse level of the fault is low with respect to the gearing in a machine. A failure mechanism that is causing low-velocity or acceleration signals from an accelerometer may be masked by the normal vibration of another component. A spectrum for a gearbox in which gear mesh masked a bearing failure is shown in Figure 7.7. The overall levels of pulses and vibration did not change when the bearing failed, but spectrum analysis showed that a bearing failure was imminent. Thus, when random noise and vibration are present, simple screening methods may be ineffective, especially those that depend on pulses.

More elaborate methods. A more sophisticated level of screening involves band filtering; that is, displaying changes in vibration in distinct frequency bands. The spectrum shown in Figure 7.7 has been divided into six frequency bands that separate faults from mass unbalance or misalignment (1x, 2x, 3x, and 4x), bearing frequencies, and gear-mesh frequencies. In this case, a defect on a rolling element bearing was masked by the gear-mesh frequency. Note that the areas of the two gear-mesh frequencies are bands within the total bearing frequency range. Thus, overall changes indicated by a single velocity reading did not indicate an impending bearing failure, but filtered results did. Distinct frequency ranges can be screened with an electronic data collector.

Amplitude screening is often unsuccessful in detecting rolling element bearing defects. The spectrum and time waveform must be used to study the frequencies and energy. The situation may arise in monitoring some rolling element bearings in low-speed machines. The peak vibration obtained from a time waveform or peak-detection circuit may be more sensitive to bearing condition than spectral measurements. Figure 4.3 contains data from a bearing cap measurement that indicate an outer race defect. The rms velocity is 0.078 IPS; the peak velocity is 0.306 IPS. However, sidebands have not yet appeared as they have in Figure 4.17. In some types of high-speed machines (3,600 RPM and above) bearing failures occur rapidly and defects in the bearing frequency range create very low levels of vibration. Bearing defects may appear more distinctly in the high-frequency range (5 kHz to 40 kHz) of the spectrum. These high-frequency responses are typically natural frequencies excited by the bearing

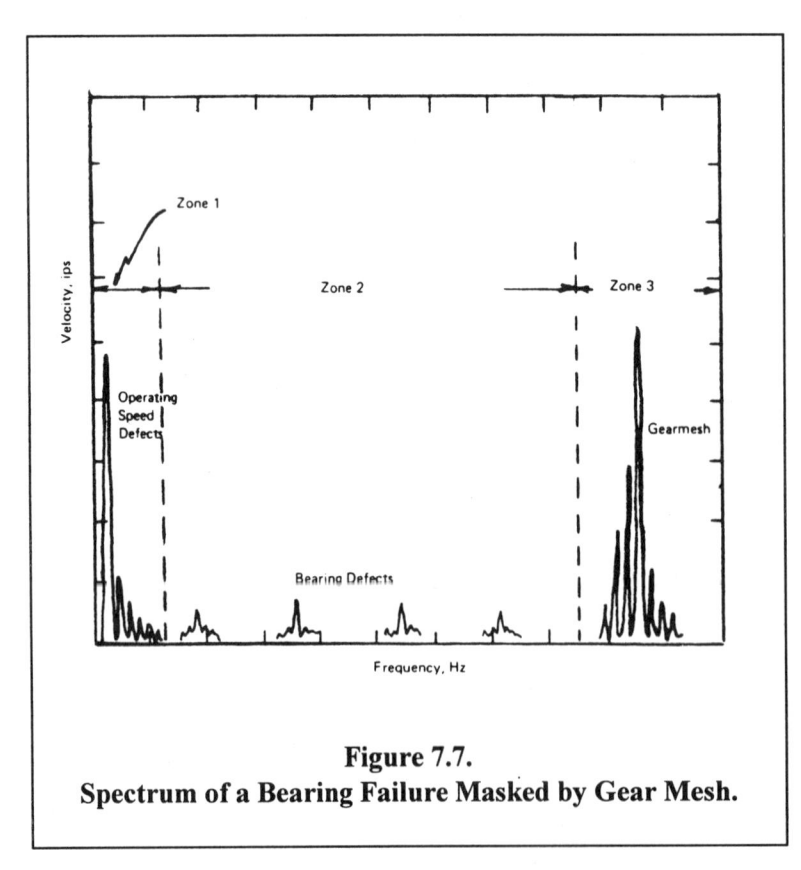

**Figure 7.7.
Spectrum of a Bearing Failure Masked by Gear Mesh.**

faults. Enveloping methods may work well on these cases. The low-frequency at operating speed and gear meshing-induced vibrations are filtered out of the signal; otherwise, their higher amplitudes create a dynamic range problem. The filtered signal is demodulated to produce a signal free of the natural frequencies. A spectrum of the demodulated signal will show the bearing frequencies and the nature of the defect.

Trending

Any vibration-related or process characteristic can be recorded for hours, days, months, or years to establish a trend. Monthly trending (Figure 7.8) is the most popular method for periodically monitoring machine condition. If masking is a problem, filtering, or band trending (filtering overall vibration and retaining only data in a frequency band — see Figure 7.7; usually data in the band are rms averaged), may be useful.

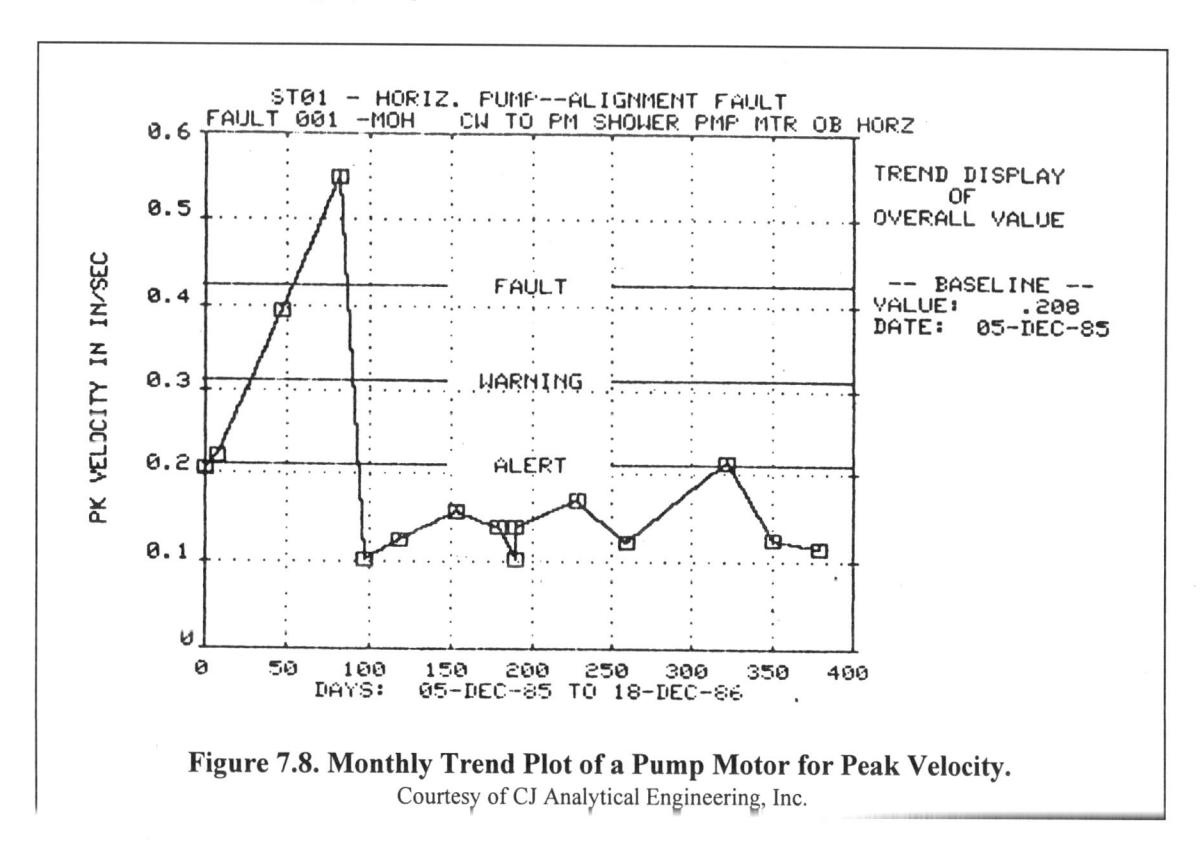

Figure 7.8. Monthly Trend Plot of a Pump Motor for Peak Velocity.
Courtesy of CJ Analytical Engineering, Inc.

Trending in several frequency ranges will provide more detailed information. It is possible to trend various measures (velocity$_{rms}$, velocity$_{peak}$, acceleration$_{peak}$, single value high-frequency acceleration) as well as process characteristics (pressure, temperature, load, and speed). Because vibration levels are often sensitive to process characteristics, it is a good idea, if possible, to normalize vibration characteristics for process conditions before trending is begun.

Alarms

Two or three alarms are typically used in the trending process. An alert alarm (see Figure 7.8) that may initiate collection of a spectrum or a time waveform when a periodic measurement is to be taken signifies the deterioration of machine condition. The trend diagram for a motor (Figure 7.8) shows alert, warning, and fault alarm levels. The alert alarm means that a detailed vibration analysis should be performed. The data usually collected as an exception (data are above alarm; see Figure 7.3) are spectrum and/or time waveform. Alarms are typically established on the basis of condition changes as indicated by a two- to two and one-half-change in machine vibration level. After the data are evaluated, either maintenance action is taken, more frequent monitoring is initiated, or regular periodic monitoring is continued. The warning alarm indicates more serious problems and should lead to a full-scale analysis or maintenance action. The time allotted for action is typically limited to a shutdown. The fault alarm means that failure is close if no maintenance action is taken. Maintenance action would include balancing, repair, redesign, or more careful installation.

Establishing realistic alarms requires knowledge of machine condition and vibration signals. In a newly established program, such knowledge is not available, and the alarms must be set based on information about other equipment, experience of others, or general vibration standards. In the vibration chart shown in *Chapter V* (see Table 5.2) the surveillance and unsuitable-for-operation levels could be used for the alert and warning alarms. If proximity probes are being used to measure shaft vibration, the ratio of the vibration level to the bearing clearance should be used to set the alarms (see Table 5.1). Alarm levels should be reviewed from time to time and changed to reflect the experience gained during the monitoring program. In this case, surveillance and shutdown limits can be used to set a three alarm system.

Example 7.1: Setting alarm levels on a motor pump unit.

Set the alarm levels for data collection on a 300 HP motor pump unit. Both units have rolling element bearings; operational speed is 1,200 RPM. Use Table 5.2 without service factors to determine the levels for rms velocity measure.

$$\text{alert} = 0.12 \text{ IPS}$$
$$\text{warning} = 0.28 \text{ IPS}$$
$$\text{shut down} = 0.6 \text{ IPS}$$

> **Example 7.2**: Setting alarm levels on an 18,000 HP turbine.
>
> The alarms are to be set on an 18,000 HP mechanical drive turbine that operates at 10,000 RPM and has bearing clearances of 8 mils. Use Table 5.1 to determine alarm values.
>
> $$\text{alert} = 0.2 \times 8 = 1.6 \text{ mils peak to peak}$$
> $$\text{warning} = 0.4 \times 8 = 3.2 \text{ mils peak to peak}$$
> $$\text{shutdown} = 0.6 \times 8 = 4.8 \text{ mils peak to peak}$$

Reports

Every periodic monitoring program should be capable of generating reports that keep management informed and accumulate technical data that result in a more efficient program. Data storage and retrieval should be thoroughly planned for accessibility and utility. Data compression techniques are useful in minimizing the space required for storage of data in the long term.

Reports on each route used in routine data collection should, at the minimum, contain a last measurement report (see Figure 7.3). Included should be a description of the measurement point, measures, date of measurement, previous value, last value, percentage of change, alarm status, trend plot for each data point, spectral data on exceptions and alarms, recommendations for maintenance action, and an executive summary (*see Chapter VI*).

Summary of Periodic Monitoring

- Periodic monitoring is used to assess the condition and changes in the condition of machines.
- Measurements are selected that provide the greatest sensitivity to a change in machine condition with the least complexity and data processing.
- For a new periodic monitoring program, machines should be listed and placed in a hierarchical order of importance to production.
- Equipment knowledge paves the way for accurate machine fault and condition analysis and should be consolidated in one table.
- Data collection routes are based on plant layout, machine train, machine type, or data type.
- Measures and measurement points are selected for efficient collection of data pertaining to condition; redundant measurements should be eliminated as experience is gained with the program (*see Chapter II*).
- Frequency spans used in measurement are based on the frequencies of the machine. (*see Chapter II*).
- Baseline data provide a reference for evaluating changes in condition.

- The frequency of data collection is based on mean time to failure of machine components, costs of failure, available personnel, number of spares, and monitoring costs.
- Chronic problems dilute the resources for monitoring good machines and should be solved.
- Screening can be a low-cost method for detecting changes in machine condition.
- Band screening may be necessary to obtain the sensitivity required to assess changes in machine condition in complex machines with rolling element bearings.
- Trending provides the opportunity to compare screening measures and the alarm settings that either initiate analysis, more frequent monitoring, or repair.
- Alarms are used in periodic monitoring to bring to the attention of the data collector that a significant change in condition has occurred.
- Two or three alarm levels are typically established on the basis of a two or two and one-half to one increase in the trended measure.
- The effect of process changes on a trended measure must be taken into account during trending.
- Reporting is used to document case histories, record alarms, and request maintenance action.
- Report formats should report the facts in a simple manner to the proper authority.
- Reports should include case histories of unusual problems, information about pre- and post-overhaul conditions, alarms and warnings, and information about time and materials and cost accounting.
- Baseline data provide a reference for evaluating changes in condition.

References

7.1. Ehrich, F.F., *Handbook of Rotor Dynamics*, 2nd ed, p 4.68, Krieger Pub. (1998).

7.2. Mitchell, John S., *Introduction to Machinery Analysis and Monitoring*, 2nd ed, PennWell Books, Tulsa, OK (1993).

CHAPTER VIII
BASIC BALANCING of ROTATING MACHINERY

Force reduction provides direct vibration control.

The forces on the bearings, structure, shaft, and couplings resulting from mass unbalance[1] are unacceptable because they sometimes will lead to vibrations that cause premature failures, unacceptable noise, and general discomfort to those around the machine. Mass unbalance occurs in a rotating machine when the center of the mass does not coincide with the geometric center (Figure 8.1). The result is a heavy spot. A small balance weight (Figure 8.2) is positioned opposite the heavy spot to reduce the mass unbalance forces.

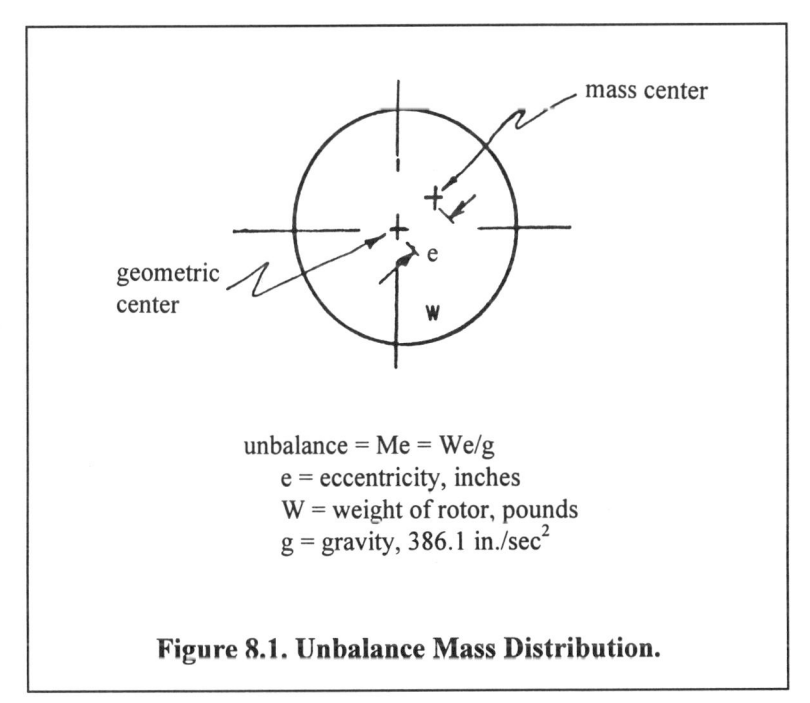

unbalance = Me = We/g
e = eccentricity, inches
W = weight of rotor, pounds
g = gravity, 386.1 in./sec^2

Figure 8.1. Unbalance Mass Distribution.

A simple test to determine the gross mass unbalance in a rotor is to set the journals on knife edges. If the rotor rolls to the same position each time it is angularly repositioned, that position is the heavy spot. Some causes of mass unbalance are listed in Table 8.1.

The amount and position of the unbalance in any machine are in general unknown, and the proper correction in the selected correction planes must be established by tests. In 1934 E.L. Thearle [8.1] outlined a procedure for two-plane balancing of rotating machines in their operating environment. The same procedures are used today. Vibration measurements have improved, and programmable calculators and data collectors are commonly used in a black box approach to balancing. This chapter describes single-plane balancing using the vector method, types of unbalance, prebalancing checks, the equipment required for balancing, vibration measurements, terminology, balancing pitfalls, selection of trial weights, and balance quality.

---

[1] Terminology of the International Standards Organization Technical Committee 108.

Figure 8.2. Balance Weight.

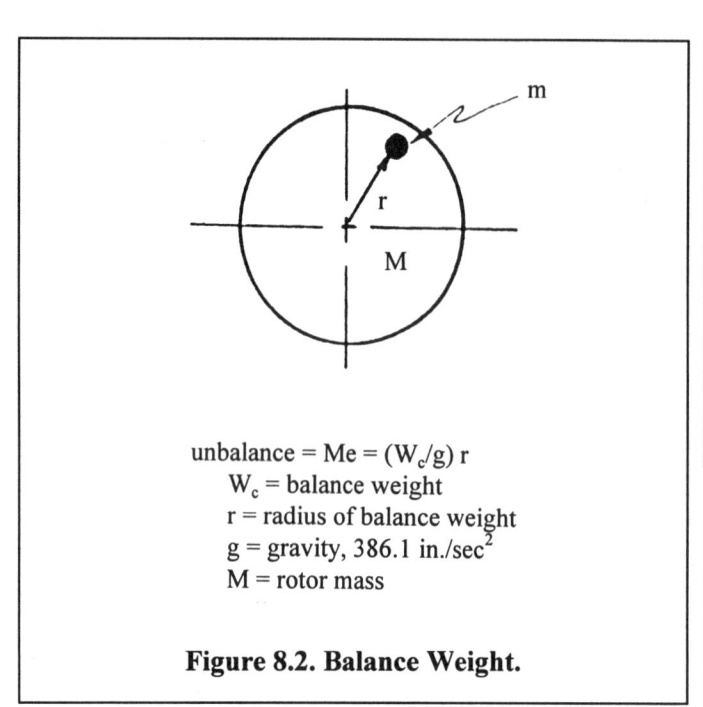

unbalance = Me = $(W_c/g)\, r$
W_c = balance weight
r = radius of balance weight
g = gravity, 386.1 in./sec^2
M = rotor mass

Table 8.1. Causes of Unbalance.

eccentricity
casting blow holes
keys and keyways
mechanical distortion
thermal distortion
corrosion and wear
deposit buildup
unsymmetrical design
component shift (motors, fans)

Table 8.2. Types of Unbalance.

static and couple — Figure 8.3
overhung dynamic — Figure 8.4
dynamic — Figure 8.5

Types of Unbalance

The major types of unbalance are classified in Table 8.2. Figure 8.3 shows pure static and pure couple unbalance for a rigid rotor. The magnitude of the unbalance is dependent on the location of the center of the mass (designated by the distance between the mass center and geometric center — sometimes called eccentricity), the total mass, and the square of the speed.

$$F = me\omega^2$$

F = force due to mass unbalance, pounds
$m = W/g$, mass of rotor or component
e = eccentricity, inches
ω = machine speed, radians/second
$\omega = 2\pi N/60$
N = machine speed, RPM

For rigid rotors the position of the eccentricity does not change with speed.

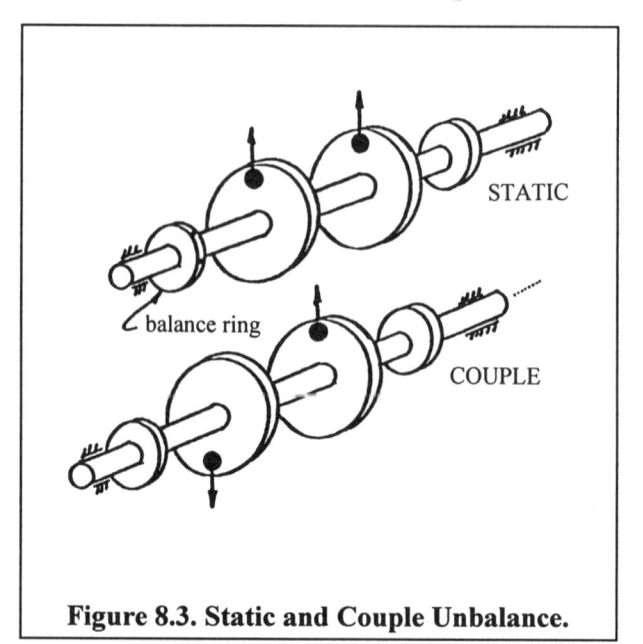

Figure 8.3. Static and Couple Unbalance.

The static unbalance can apply to rotors that can be balanced with one or two planes. Couple unbalance is usually associated with two-plane rotors; however, overhung fan rotors (Figure 8.4) often exhibit couple or dynamic combination static and couple unbalance; neither is easy to correct with single-plane balancing techniques.

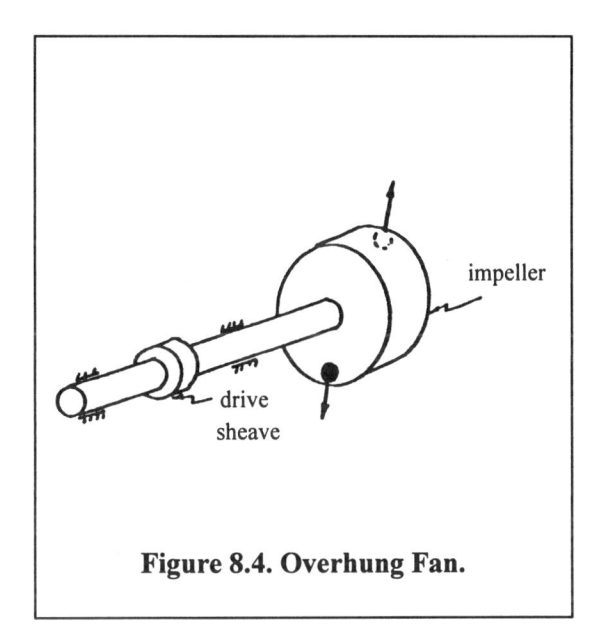

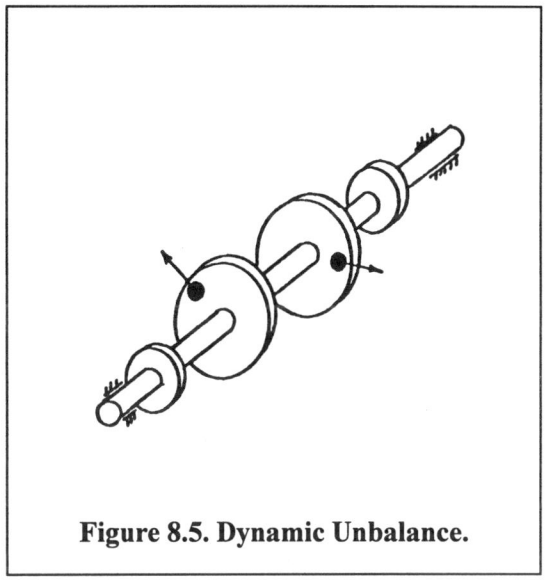

Figure 8.4. Overhung Fan.

Figure 8.5. Dynamic Unbalance.

The combination of static and couple unbalance is termed dynamic unbalance (Figure 8.5). Either two-plane techniques or trial-and-error procedures are required for correction.

Balancing Equipment

Table 8.3 lists the equipment required for general balancing. The meter must have a synchronous speed-tracking filter so the operating-speed component of vibration associated with the mass unbalance can be obtained. The amplitude of vibration, as well as the phase angle between a reference point on the rotor (e.g., a keyway) and the peak vibration, are obtained. Trial weights of a size and shape appropriate for the machine should be available. The equipment listed in Table 8.3 can be used to calculate the correction weight, but most data collectors are able to perform this function. It is a good idea to know the vector method, however.

**Table 8.3.
Equipment Required for Balancing.**

meter that reads phase and amplitude
electronic calculator
vector addition/subtraction program
two-plane balancing program
polar graph paper
trial weights
protractor
rule with tenths scale
parallel rule or triangles

Prebalancing Checks

Before beginning to balance, other possibilities for the problem should be eliminated. A complete vibration analysis should be carried out to assure that mass unbalance is the problem. Table 8.4 lists some routine checks and analyses that will be helpful in isolating the problem and assessing the nature of the equipment to be balanced.

Table 8.4. Prebalancing Checks.

nature of unbalance problem (do a vibration analysis)
determine whether or not the rotor is clean
assess rotor stability (structural, thermal)
determine critical speeds (start-up/coast-down tests)
locate balance weights already in place
know details of balance planes or rings

If unbalance is not the problem, correct whatever is wrong — e.g., excessive bearing clearance, looseness, resonance, misalignment — before attempting to balance. Otherwise, the procedure will fail. If the rotor is not clean and material breaks away during or after balancing, the results will be unsatisfactory. If the rotor is not stable as a result of structural deflection, thermal distortion, beating, or operation close to a critical speed, the phase angle will drift and shift. Phase angle readings must be accurate to 15° if there is to be any improvement. Accurate phase readings are important in balancing. Start-up and coast-down tests are used to determine critical speeds and resonances; attempts to balance at these speeds may result in unstable phase angle readings. It is important to know the diameters of balance rings in order to determine applied forces.

Measurements

Sensors that are sensitive to balance should be selected (Table 8.5). Proximity probes provide the most direct measure. However, runout must be subtracted. Velocity transducers and accelerometers mounted on casings are indirect measures of vibration. Either the photoelectric sensor or the proximity probe provides the most accurate phase readings. A strobe light provides physical insight into shaft behavior; strobe light phase conventions are listed in Table 8.6.

Table 8.5. Sensors.

For vibration:
proximity probes
velocity transducers
accelerometers

For phase:
strobe light
photoelectric sensor
proximity probe

**Table 8.6.
Strobe Light Conventions.**

1. Stationary protractor – numbered with rotation – positive phase angle in direction of rotation

2. Rotating protractor – numbered with/against rotation – positive phase angles with/against direction of rotation

Figure 8.6 shows proximity probe measurements of phase angles and vibration levels. Measurements made with a strobe light and a velocity transducer are given in Figure 8.7. The proximity probe measurement and triggering signal have no electronic lag. This means that the probe measurement leads to the phase angle between the sensor and the high spot directly. The velocity transducer and strobe light have an electronic phase lag (see Figure 8.7). Both systems have a mechanical phase lag between the high spot (peak vibration) and heavy spot (location of mass unbalance); that is, they are instrument dependent. Displacement is the preferred measure for balancing even when velocity transducers are used.

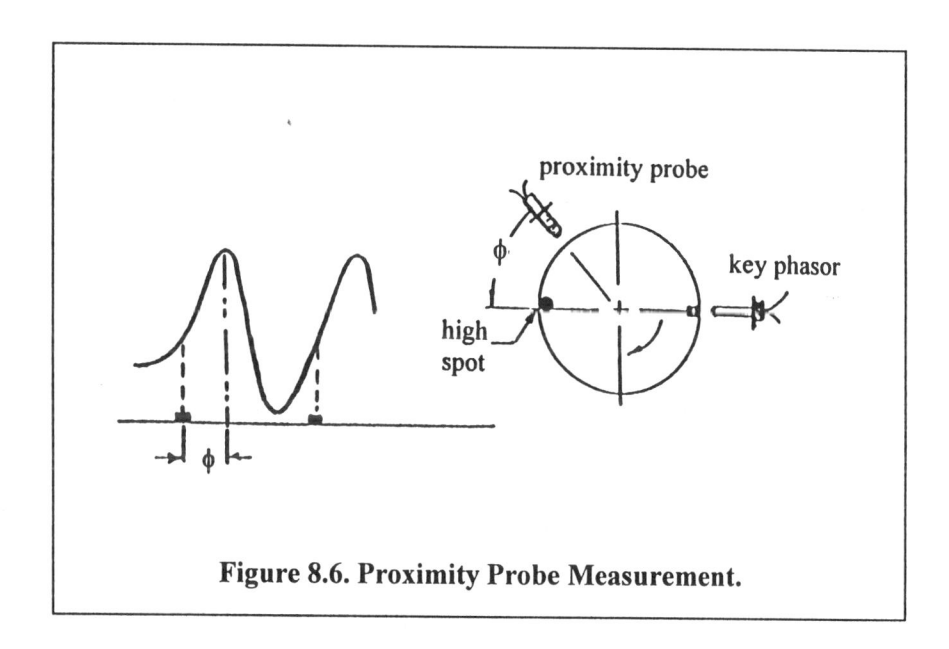

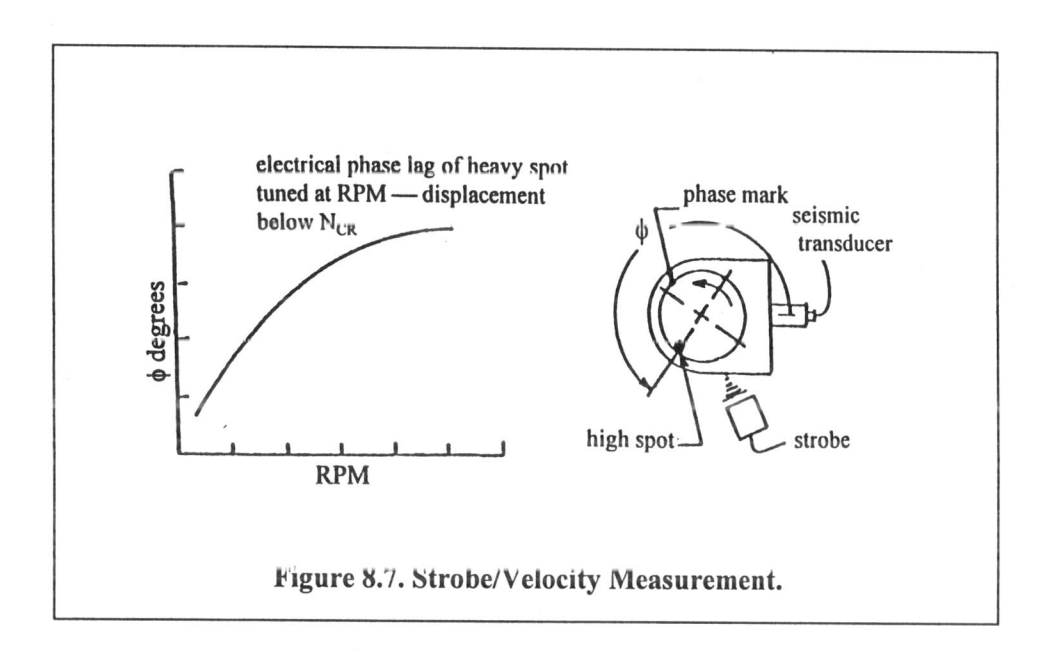

Figure 8.6. Proximity Probe Measurement.

Figure 8.7. Strobe/Velocity Measurement.

Relationship Between Mass Unbalance and Phase

The angular location of mass unbalance on the rotor is determined from a known mark (see Figure 8.6). This mass unbalance (heavy spot) generates a force that leads the vibration peak (high

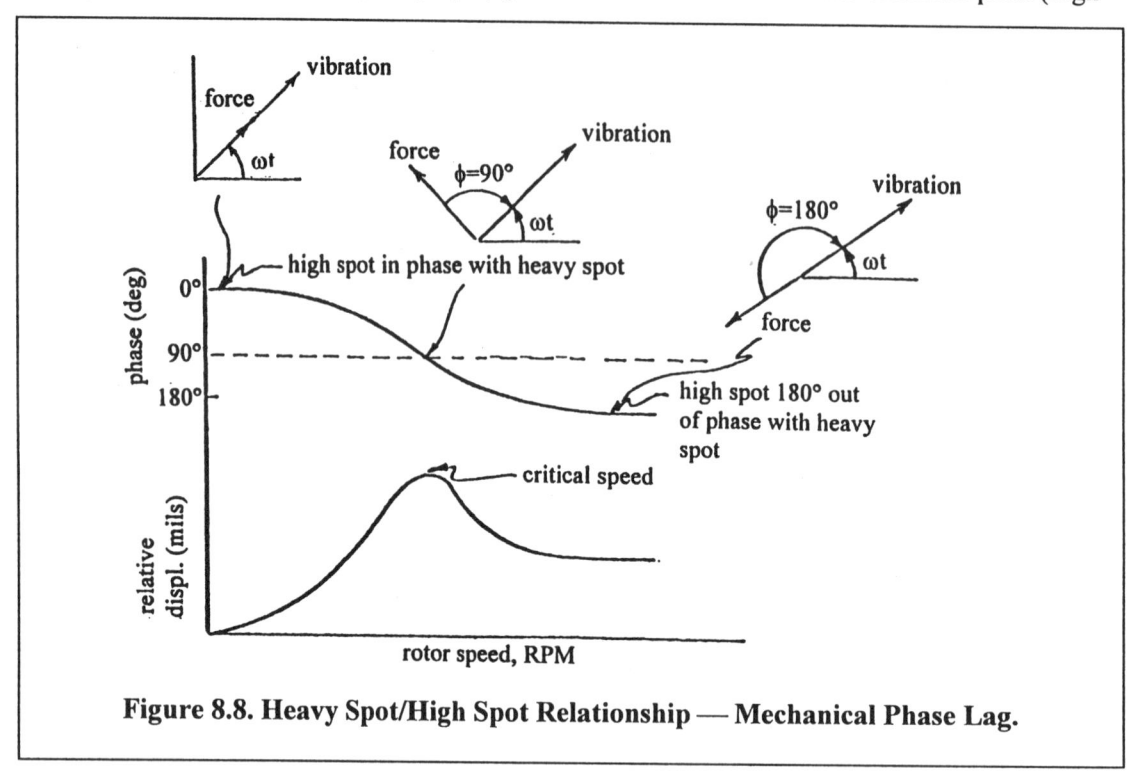

Figure 8.8. Heavy Spot/High Spot Relationship — Mechanical Phase Lag.

spot) by 0° to 180°, depending on the location of the operating speed with respect to the critical speed. This is termed the mechanical phase lag; that is, vibration lags the force that causes it (Figure 8.8).

When the location of a trial weight is selected, the peak vibration can be related to the heavy spot on the shaft if the mechanical and electronic phase lags are known. The electronic phase lag is provided by the manufacturer of the vibration measuring instrumentation being used. Proximity probes, accelerometers, and photoelectric transducers have no electronic phase lag.

The location of the operating speed relative to the critical speed, a measure of the mechanical phase lag, can be obtained by measuring amplitude and phase during a coast-down test. Of course, if the critical speed of the machine is known, this test need not be done. A rotor operating at a speed of less than 50% of the critical speed is in a rigid mode, and the heavy spot is close to the measured high spot and is altered only slightly by damping (see Figure 8.8). As the first critical speed is approached, the high spot lags the heavy spot more and more, to a maximum of 90°

at the critical speed. After the first critical speed is passed, the phase lag increases until it reaches 180°. Therefore, the trial weight is placed opposite the measured high spot if the operating speed is well below the critical speed. Place the trial weight on the measured high spot if the operating speed is well above the critical speed.

Trial Weight Selection

Proper selection of a trial weight may save time as well as a machine. Jackson [5.2] has suggested that a trial weight yielding a force of not more than one tenth (10%) the static weight of the machine rotor be used. The trial weight can be calculated from the formula. The smallest possible trial weight should be used. If no vibration response is obtained, either the trial weight is too small or the problem is not mass unbalance.

$$W_T = 56{,}375.5\ (W/N^2 e)$$
W_T = trial weight, ounces
e = eccentricity of trial weight, inches
W = static weight of rotor, pounds
N = rotor speed, RPM

Balancing Pitfalls

Table 8.7 lists a number of pitfalls in the balancing process. Some of them are obvious. Errors in data are most common when thermal sensitivity is a problem. If there are thermal problems, tests involving hours of machine operation may be necessary to obtain acceptable data. Trade-offs for specific machine conditions may be needed.

Table 8.7. Balancing Pitfalls.

balancing syndrome
error in original
balancing at low speeds
inaccurate data
thermal sensitivity
dirty rotor
resonance and critical speeds
loose rotor
loose supports
removal of trial weight

Vector Method with Trial Weight

A single-plane balancing procedure is summarized in Table 8.8. The motor is operated at a selected speed and the amplitude and phase are measured. The high spot (*a* in Figure 8.9) is marked and its amplitude *oa* is laid out to scale. Damping, stiffness, and mass will cause the phase of the vector *oa* to lag the unknown position of the heavy spot.

A trial weight W_T is placed at a selected position, the rotor is operated at the same speed as before, and a new high spot *b* is identified. The new amplitude vector *ob* represents the effects of both the original unbalance and the added trial weight W_T. The vector difference *ab* = *ob* - *oa* is the effect of W_T alone.

Table 8.8. Vector Method with Trial Weight.

measure and record signal
install trial weight
measure and record trial run
calculate vectors
correct trial weight
measure and record trial run

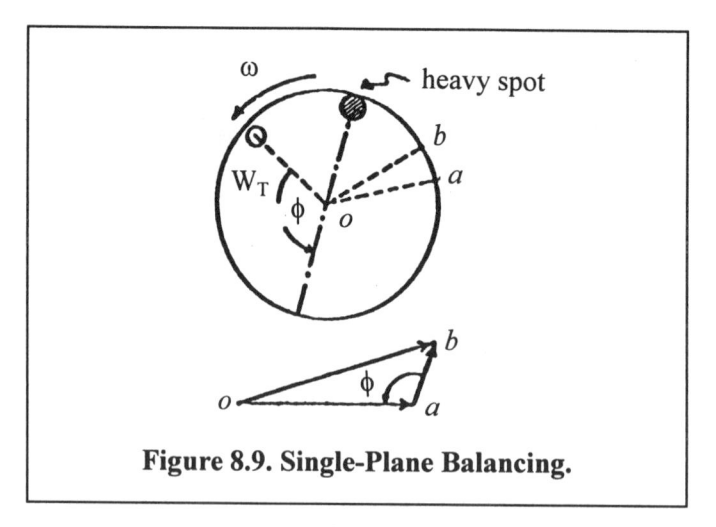

Figure 8.9. Single-Plane Balancing.

Move the W_T in the same direction and angle φ as required to make *ab* parallel to and opposite *oa*. The trial weight is increased or decreased in the ratio *oa/ab* to equal the original unbalance. If *ab* is smaller than *oa*, the trial weight should be increased. The procedure is shown in Figure 8.10. Table 8.9 is a procedure for using a vector diagram for single-plane balancing.

Table 8.9. Procedure for Constructing a Vector Diagram for Single-Plane Balancing.

1. Mark the direction of rotor rotation on the graph.
2. Mark the direction of positive phase angle.
3. Establish a scale of numbers of mils per division so the vectors are large but do not exceed the graph.
4. The original vibration O (5 mils at 190° in Figure 8.10) is plotted on the graph.
5. The location of the trial weight (W_T) is plotted (30°) and its size (75 grams) are noted on the graph.
6. Plot the vibration (O + T) obtained after the trial weight has been added to the rotor. The rotor must be operated at the same speed as when the original data (O) were acquired.
7. The difference between (O) and (O + T) is the effect of the trial weight.
8. The effect of the trial weight is obtained by drawing a line between (O) and (O + T).
9. (O) + (T) must be equal to (O + T). Therefore, the arrow on (T) must point to (O + T). Vectors add heads to tails and subtract heads to heads.
10. (T) is now repositioned with its tail at the origin by moving it parallel and maintaining the same length.
11. Draw a line opposite (O) from the origin.
12. The goal in balancing is to add a trial weight that will create a (T) vector directly opposite and equal to (O).
13. The angle between (T) and the line opposite (O) 36° determines how far and in what direction the trial weight must be moved (see Figure 8.10).
14. The trial weight is multiplied by the ratio of the original vibration to the effect of the trial weight (5/3.4) to determine the balance weight. 75 g (5/3.4) = 110 g.

Weight Splitting and Consolidation

Figure 8.11 and Figure 8.12 are examples of weight splitting and combination. The desired locations for the weights a and b are selected. The actual position of the weight and its magnitude are marked on the polar plot (see Figure 8.11). A parallel rule is used to determine graphically the magnitudes of the weights at a and b by the lengths of the vectors. Weight combination is the inverse procedure (see Figure 8.12).

Acceptable Vibration Levels

For field balancing, Figure 5.4 is used to obtain acceptable residual vibration levels on the bearing cap resulting from mass unbalance. Figure 5.2 is used in the field for proximity probe measurements.

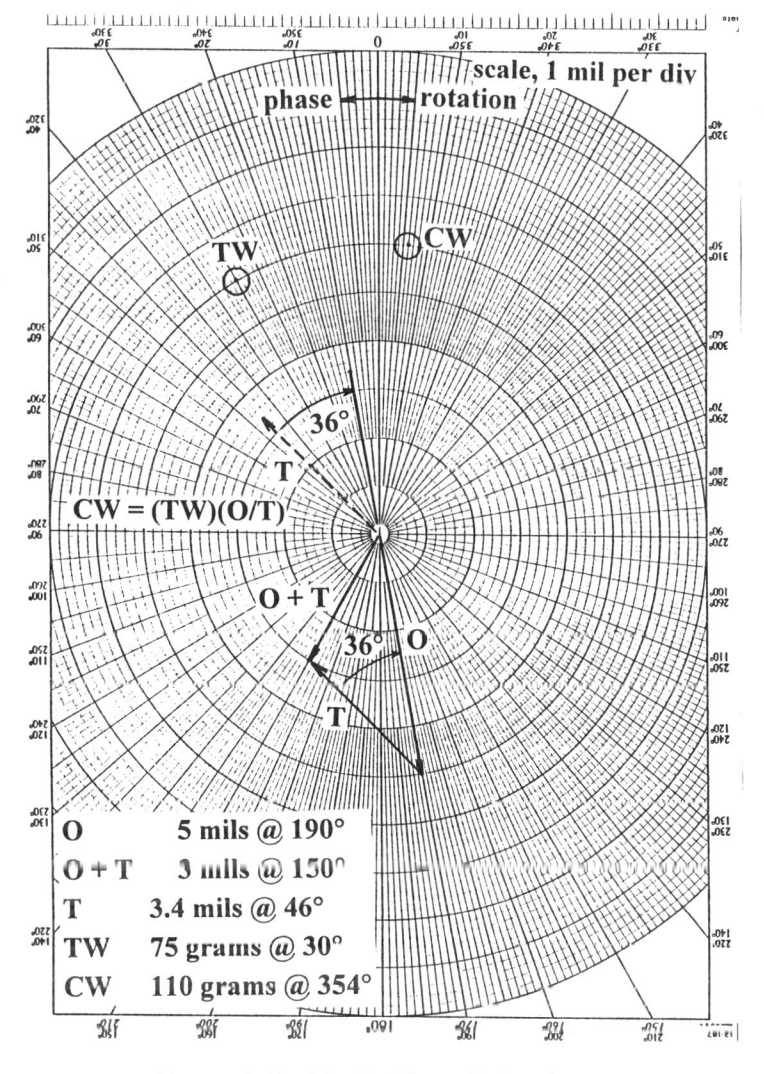

Figure 8.10. Single-Plane Balancing.

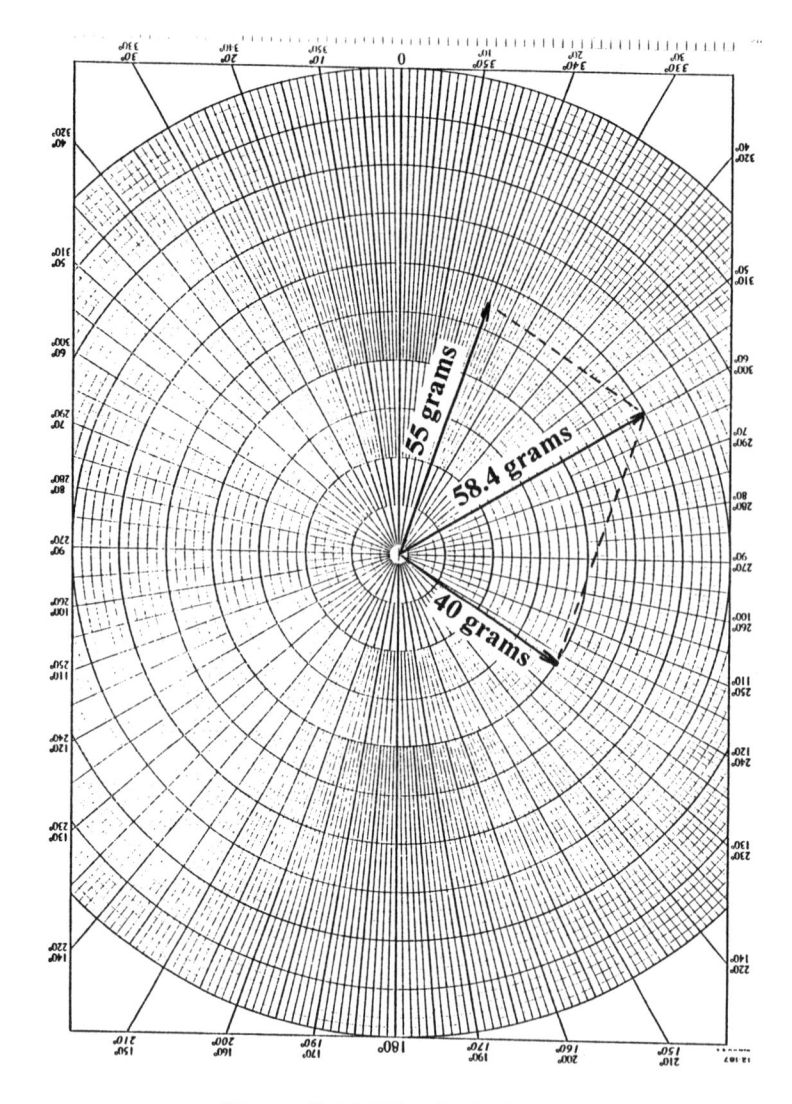

Figure 8.11. Weight Splitting.

Summary of Basic Balancing of Rotating Machinery

- Mass unbalance of a rotor results when the mass center is not at the same location as the geometric center.
- Mass unbalance causes a rotating force at the frequency of shaft speed.
- The amount of mass unbalance force depends on the location of the mass center from the geometric center, the weight of the object, and the square of the speed.
- Balancing is a procedure in which a balance weight that creates a force equal to the mass unbalance is placed opposite the effective location of the mass unbalance.
- The heavy spot is the angular location of the mass unbalance on the rotor.
- The high spot is the angular location of the peak of vibration (displacement).

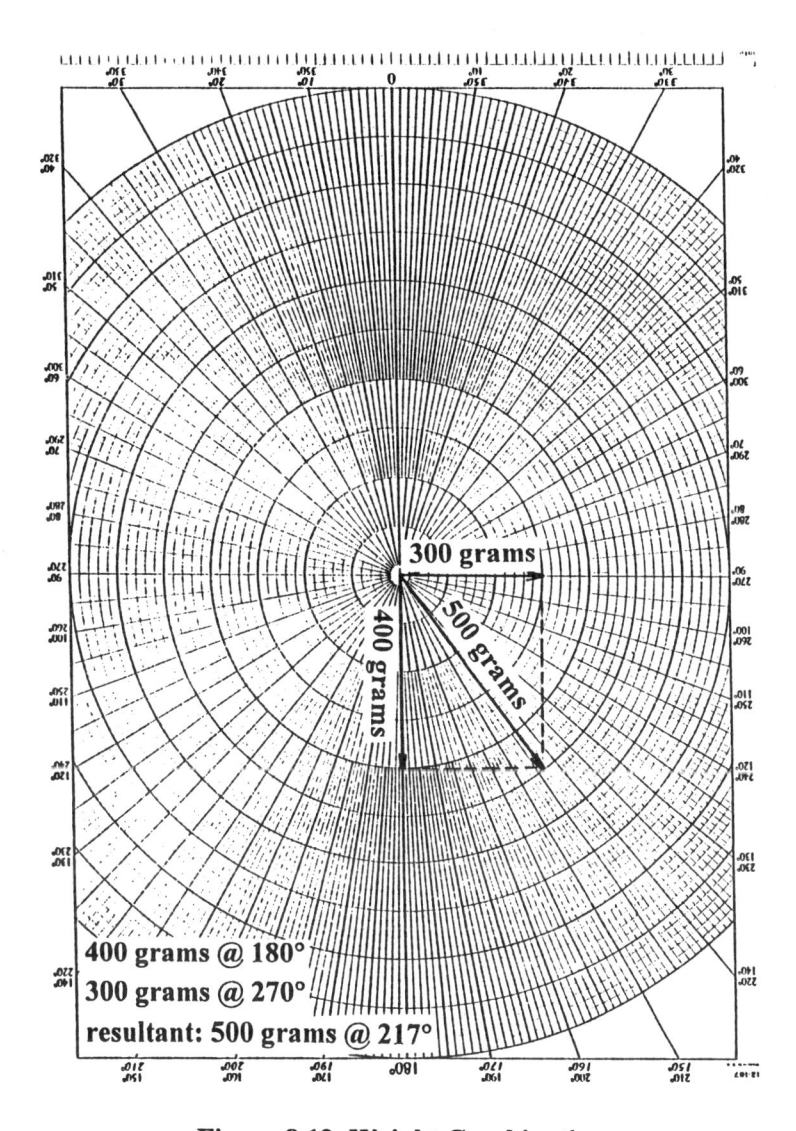

Figure 8.12. Weight Combination.

- The high spot is measured during the balancing process; however, the balance weight must be positioned opposite the heavy spot. Either displacement, velocity, or acceleration can be measured; however, displacement is preferred.
- The high spot lags the heavy spot as a result of electronic (instrument) and mechanical lag.
- Balancing should not be performed until it is evident that misalignment, excessive bearing clearance, looseness, and distortion are not the cause of the vibration at operating speed.
- The rotor should be clean and structurally sound prior to balancing.
- Trial or calibration weights are used to obtain the mechanical lag.
- The rule of thumb for selecting a trial weight is that it should create a force of no more than 10% of the rotor weight.
- The vector method is used to determine the size and location of the correction weight.

- Vibration is measured on the machine with and without the trial weight. The vectorial difference is determined to assess the effect of the trial weight. The trial weight is moved relative to the effect vector so that it is opposite the original unbalance vector. The size of the trial weight is adjusted so that the effect vector is the same length as the original unbalance vector.
- Allowable field unbalance values are obtained from vibration severity levels in ISO 2372 (rms), found in Table 5.3, and the Blake chart (peak), found in Figure 5.4.

References

8.1. Thearle, E.L., "Dynamic Balancing of Rotating Machinery in the Field," Trans. ASME, **56**, pp 745-753 (Oct 1934).

8.2. Eisenmann, Sr. R.C. and Eisenmann, Jr. R.C., *Machinery Malfunction Diagnosis and Correction*, Prentice Hall PRT (1998).

INDEX

acceleration, 1.1; 1.7; 2.1
accelerometer, see transducer, accelerometer
acceptance, levels, 5.7
acceptance, machine, 6.1
acceptance tests, 6.4
alarms, 7.8; 7.16; 7.17
aliasing, 3.7
amplifier, horizontal, see time base
amplifier, vertical, 3.2
amplitude, 1.2; 1.5
amplitude, calculation, 1.10
amplitude, effective, 5.7
amplitude, peak, 1.4
amplitude, peak to peak, 1.4
amplitude, rms, 1.4
analog integrator, 2.8
analysis, spectrum, see spectrum analysis
ANSI S2.41, 1985 (R 1990), 5.9
API 670, 2.16
API 678, 2.16
autoranging, 3.17
averaging, 3.12
averaging, overlap, 3.12
averaging, peak hold, 3.12
averaging, rms, 3.12
averaging, synchronous time, 3.12
balancing, 8.1
balancing pitfalls, 8.7
balancing, single-plane, 8.7
bands, frequency, 7.14
baseline data, 6.1; 7.11
baseline tests, 6.4
Baxter, N.L., 4.41
bearings, analysis techniques, 4.16
bearing, angular contact, 2.13
bearings, defects, 4.16
bearing, excessive clearance, 4.19
bearing, frequencies, 4.14; 4.16
bearings, measurement techniques, 2.13; 4.15
bearings, inner race defect, 4.18
bearings, rolling element, 4.14
bearings, sidebands, 4.15; 4.16; 7.14
bearing vibration, 5.3
beats, 4.2; 4.4
bent shaft, see rotor bow
bins, see lines
Blake Chart, 5.6
Campbell, W.R., 4.41
compressors, 4.30; 4.39
Crawford, A.R., 2.16

critical speeds, 1.15; 4.6; 4.8
critical speed tests, 6.5; 6.8
damping, 1.2; 1.3
data acquisition, 2.1; 6.3
data acquisition plan, 6.1
data acquisition time, 3.3; 3.7
data collection, frequency of, 7.12
data, presentation of, 6.12
data processing, 3.1
data sampling, 3.5
defects, operating speed, 4.7
deflection, 1.3
design faults, 4.1
design verification, 6.1
displacement, 1.1; 1.6; 2.1
distortion, 4.12
dynamic range, 3.11; 3.16
eccentricity, 4.11; 4.14
eccentricity ratio, 5.2
environment, 6.12
excessive clearance, see wear, bearing
excitation, 1.1; 6.6
electronic data collectors, 3.4; 3.5
fans, 4.30; 4.36
fans, blade-pass frequency, 4.37
fans, faults, 4.37
fan, characteristics at constant speed, 4.36
FFT algorithm, 3.8
FFT analyzer, 1.13; 3.3
fault diagnosis, 4.1; 6.1
faults, operating speed, 4.6
filters, anti-aliasing, 3.8
fluid-film bearing, 1.3
forces, vibratory, 1.2
format, in data presentation, 4.6
formulas, conversion, 1.9
frequency, 1.1; 1.2
frequency, forcing, 1.15
frequency, harmonic, 1.4; 1.7
frequency, lowest resolvable, 3.6
frequency, magnitude, 1.1
frequency, monitoring, 7.12
frequency, natural, 1.15; 6.5
frequency, response, 2.4
frequency spans, 2.13
gearboxes, 4.20
gearboxes, malfunctions, 4.21
gear, broken tooth, 4.23
gear, eccentric pinion, 4.23
gear mesh, frequencies, 4.21
gear mesh, problems, 4.21
gears, 2.13

gears, axial measurements, 2.13; 4.22
gears, spur, 4.22
gears, malfunctions, 4.21
gravitational constant, 1.3
HFD methods, 4.19
harmonics, 4.16
heavy spot, 8.1; 8.6
high spot, 8.5; 8.6
impact test, 6.7
input, external intensity, 3.2
installation, 4.1
interference diagrams, 4.7; 6.6
ISO 2372, 1974, 5.9
ISO 7919, 1986, 5.9
Jackson, C., 5.9
limits, casing, 5.5
lines, 3.3; 3.5; 3.6
load zone, 2.12
looseness, 4.11; 4.13, 4.7
machine, condition evaluation, 5.1; 6.1
machine, excitation, 7.4
machine, excitation and response, 7.3
machine, ratio of casing to rotor weights, 5.2
machine testing, 6.1
machine, listing and categorization, 7.2
machinery knowledge, 7.2
machines, classification for monitoring, 7.2
Maedel, P.H. Jr., 5.9
magnetic pickup, 2.9
malfunction, 1.1
masking, 7.14
mass, 1.1
mass unbalance, 4.9; 8.1
mass unbalance, forces, 8.2
measurement, axial, 2.13
measurement, casing, see vibration, acceptable levels
measurement points, 6.2; 7.8
measurement report, 7.8
measurement, shaft, see vibration acceptable levels
measure(s), 1.6; 2.1; 2.3; 5.1; 7.8; 7.9
measures, conversion between, 1.7
misalignment, 4.10
Mitchell, J.S., 7.18
mode shape, 1.16; 6.5
monitoring, periodic, 7.1
motion, harmonic, 1.4

i

motion, periodic, 1.4
motor, broken rotor bars, 4.27
motor, magnetic center, 4.27
motor, malfunction, 4.27
motor, stator shorts, 4.26
motors, air-gap variation, 4.25
motors, electric, 4.24
motors, line frequency, 4.24
motors, synchronous motor
    speed, 4.25
mounting, 6.12
noise, 1.1
Nyquist criterion, 3.7
orbit, 2.15
orders, 1.2   *Appendix 4.1*
oscilloscope, 3.1
peak, derived, 1.6
peak hold, 3.12; 6.9
peak, overall, 1.5
peak to peak, 1.5
period, 1.2
periodic monitoring, 6.1; 7.1
phase, 1.1; 1.2; 4.1
phase angle, 1.12
phase lag, mechanical, 8.6
pickup,
    see transducer, magnetic
    pickup
polar plot, 6.10
prebalancing checks, 8.4
proximity probe,
    see transducer, proximity
    probe
pumps, 4.30; 4.31
pumps, cavitation, 4.34
pumps, recirculation, 4.33
pumps, vertical, 4.32
purchase specifications, 6.4
reports, 6.14; 7.17
resolution, 3.3; 3.14
resonance, 1.15; 4.5; 4.11; 6.5
resonance testing, 6.5; 6.7
rotor/bearing vibration,
    evaluation of, 5.4
rotor bow, 4.10
rotor position, 5.1
route mean square, rms, 5.8; 6.5
route selection and definition,
    7.5
routes, 7.7
rubs, 4.11
samples, number of, 3.5
scale factor, 1.10; 2.6
screening, 7.13
screening, band filtering, 7.14
sensor, photoelectric, 8.4
sensor, see transducer
service factor, 5.7
setup, monitoring, 7.9
severity, 5.8

shaft rider, 1.10
sidebands, 4.2; 6.14; 7.14
sidebands, bearing,
    see bearing sidebands
site inspection, 6.4 *Soft foot A5/4.7*
specifications, 6.11
spectrum, 1.14; 2.14
spectrum analysis, 4.2
speed, 1.1
stiffness, 1.1; 1.2; 1.3
strobe light, 1.12; 2.9; 8.4
structural damage, 1.1
survey request form, 7.6   *Sync. Motor Speed (SMS) 4.25*
tape recorder, 6.2
Taylor, J.I., 4.41
test equipment, selection of,
    6.3; 7.12
test plans, 6.1
Thearle, E.L., 8.1, 8.11
time base, 3.2
time waveform, 2.14
transducer(s), 1.10; 2.1; 2.4
transducer, accelerometer, 1.10;
    2.7
transducer, force, 2.7
transducer, location, 2.11
transducer, mounting, 2.11
transducer, proximity probe,
    1.10; 2.6
transducer, selection, 2.4; 2.10
transducer, velocity, 1.10; 2.6
transient data, 6.8; 7.11
trending, 7.15
trial weight, 8.7
trial weight selection, 8.7
triggering, 3.1
triggering devices, 2.8; 2.9
unbalance, causes, 8.2
unbalance, types, 8.2
velocity, 1.1; 1.6; 2.1
velocity transducer,
    see transducer, velocity
vibration, acceptable levels, 5.2;
    5.7
vibration, casing, 5.5
vibration, cause, 1.2
vibration, evaluating shaft, 5.3
vibration, guidelines for
    condition evaluation, 5.5
vibration, measurement, 1.10
vibration, shaft, 5.2
vibration units, 1.1
volt, 1.10; 1.2; 2.1
waveform, periodic, 1.13
waveform, time, 1.14
wear, 1.1
wear, bearing, 4.11; 4.13
wear, machine, 4.1
weight, 1.3
weight consolidation, 8.9

weight splitting, 8.9
window, 3.8
window, flat top, 3.9
window, FFT selection, 3.9
window, Hanning, 3.8; 3.9
window, uniform, 3.9
zoom, 3.3

APPENDIX

Terminology

acceleration	time rate of change of velocity; proportional to force acting on a body; a measure used above 1,000 Hz; measured in g's
alarm	a value of a measure used to assess a change in machine condition
aliasing	a false component in the spectrum that results when a digital sampling rate is less than two times the frequency of the data
amplitude modulation	change in vibration amplitude with time; causes sidebands in the spectrum around modulated frequency
amplitude, peak	maximum vibration value, positive or negative, in a data sample or maximum value of a spectral component
anti-aliasing filter	a low-pass filter set to eliminate frequencies that would cause aliasing
autoranging	selection of an optimum dynamic range by an instrument based on measured data
averaging, rms	rms is calculated by squaring all components of the average, adding the squares, dividing by the number of measurements (mean) and taking the square root. It does not eliminate noise but provides a good estimate of the signal plus the noise
averaging, synchronous time	with a trigger, it eliminates all noise and nonsynchronous data
bands, frequency	distinct spectral frequency spans used to trend vibration data; excludes unwanted frequency components
bandwidth	a spacing between frequencies that is calculated from the frequency span (F_{max}), number of lines (N), and window factor (WF); BW = (F_{max}/N) (WF)
beats	summation of two vibration signals of slightly different frequencies
best efficiency point (BEP)	the most efficient pump operating point on a head versus flow curve
bins	see lines
baseline	recorded reference data used for assessing changes in machine condition
Bodé plot	a plot of amplitude and phase of a signal (usual synchronous to speed) versus shaft speed
clearance, bearing	the difference between the bearing inner diameter and the journal outer diameter for a fluid-film (sleeve) bearing

Appendix

critical speed	rotor speed at which the operating speed or an order of operating speed is equal to a rotor/bearing natural frequency; a resonance caused by the rotor
damping	a vibration control mechanism that generates heat
data acquisition time	time required to acquire a record for FFT processing; depends directly on the number of lines and inversely on the selection of F_{max}
displacement	vibratory motion of a rotor or structure; measured in mils peak to peak
dynamic range	digital instrument voltage range that permits the resolution of small-amplitude signals in the presence of large amplitudes; dependent on the number of bits in the instrument
eccentricity	journal not operating on center; rotational axis off center yields a forced sinusoidal motion. The amount that a journal is not operating in the center of a fluid-film bearing
eccentricity ratio	ratio of eccentricity to radial clearance in a fluid-film (sleeve) bearing
excitation, machine	cause of mechanical vibration; e.g., mass unbalance, misalignment
FFT algorithm	calculation procedure to change a time waveform into a spectrum using sines and cosines. The number of samples, based on a power of two, is defined by the number of lines selected
frequency	repetition rate (number of cycles or events per unit time) of a periodically vibrating shaft or structure; expressed in cycles per second (Hz), cycles per minute (CPM), or orders of operating speed (e.g., 1x, 2x, 3x)
frequency, blade-pass	number of blades in a stage or row multiplied by shaft speed
frequency, forcing	frequency of a vibratory force that causes vibration
frequency, gear mesh	number of gear teeth multiplied by operating speed
frequency, natural	frequency at which a shaft or structure vibrates when subjected to a pulse or similar forcing frequency; dependent on the design of the machine
frequency response	amplitude response of an instrument or transducer to an input (excitation) as a function of frequency
frequency spans	FFT analyzer frequency settings to view a range of frequencies
frequency, vane-pass	number of vanes multiplied by operating speed
harmonics	multiples of a single frequency component in the spectrum
harmonic motion	a single frequency of vibratory motion that repeats itself over a specified time interval
heavy spot	angular location of the mass center on a rotor

Appendix

high spot	angular location of the peak vibration of a rotor
lines	display points of frequency components of the spectrum of an FFT analyzer; number of lines equals data samples (a power of 2) divided by 2.56
masking	occurs when a small-amplitude frequency component (e.g., bearing fault), which causes significant condition changes; is included in an overall measurement with high-amplitude components (e.g., gear mesh at operating speed)
mass	a measure of the resistance of a body to motion; equal to the volume of the material multiplied by its density or the weight divided by the gravitational constant, g
measure	a unit or standard of measurement that provides a means for evaluating data
mil	one thousandth of an inch
mode shape	shape of a rotor or structure when it vibrates at a natural frequency
monitoring, periodic	acquisition and trending of machine vibration data in an organized and periodic manner
motion, periodic	motion that repeats itself in a fixed time
Nyquist criterion	sampling rate must be greater than two times the maximum input frequency
operating speed	rotational speed of a machine, RPM
orbit	x-y display of rotor or pedestal position
orders	**integer multiples of sinusoidal vibration at a frequency of the machine operating speed**
overlap processing	data that are averaged without taking a completely new sample for each average after the first sample
period	time required to complete a cycle of vibration; usually measured in seconds per cycle
phase	time relationship between a vibration signal and a second vibration signal (different loacation) or a trigger signal with the same frequency
polar plot	vector plot of transient data, amplitude and phase, at varied speed
resolution	FFT analyzer setup that permits accurate display of two closely-spaced frequencies; resolution = $(F_{max}/N)(2)(WF)$
response	machine motion resulting from an excitation

Appendix

resonance	condition that results in amplification of machine vibration when the forcing frequency is close to or at a natural frequency
sensor	see transducer
severity	expression of the criticality of a measured vibration level in a machine; for ISO, vibration severity equals overall rms velocity from 10 Hz to 1,000 Hz
sidebands	frequency components that appear beside dominant direct frequencies in the spectrum; e.g., gear mesh and operating speed
speed	time rate of change of angular rotation of a shaft in revolutions per minute (RPM)
test plan	plan involving what, where, and when to measure to evaluate faults, condition, or properties of a machine
time waveform	a plot of amplitude versus time
transducer	device that converts mechanical vibration, temperature, or pressure into an electrical signal
transistor-transistor logic (TTL)	a ± 5 volt signal generated by an optical transducer for triggering vibration instruments and phase measurement
trending	periodic plotting and comparison of a monitoring measure; e.g., vibration, temperature, pressure
trial weight	small metal object applied to a rotor for calibration of balance sensitivity (ounces per mil)
triggering	initiation of data acquisition, processing, or any other action
unbalance	difference between the geometric and mass center of a rotor
vector	quantity that has amplitude and direction (phase)
velocity	time rate of change of displacement of a vibrating machine; measured in inches per second (IPS)
volt	measure of electrical potential (force)
window	mathematical function applied to a data sample to prepare it to conform to the necessary conditions for FFT transformation
zoom	method for increasing FFT resolution by reducing the frequency span and positioning the center of the span at a desired frequency location

[handwritten note: Softfoot - See Index]

Operating Speed Defects

Fault	Spectral Frequency	Page #	Spectrum, Time Waveform, Shape	Correction
critical speeds	1x, 2x, 3x, etc.	4.8	amplified vibration due to proximity of operating speed to natural frequency	tune natural frequency or change operating speed
mass unbalance	1x	4.9	distinct 1x with much lower values of 2x, 3x; vertical/horizontal 90° out of phase	field or shop balancing
misalignment	1x, 2x, occasionally 3x	4.10	distinct 1x with equal or higher values of 2x, 3x; vertical/horizontal in phase	perform hot and/or cold alignment
shaft bow	1x	4.12	dropout of vibration around critical speed in Bodé plot; 1x constant with speed	heating or peening to straighten rotor (allow rotor to float axially)
fluid-film bearing wear and excessive clearance	1x, sub-harmonics, orders	4.13	high 1x, 1/2x, sometimes 1/2 or orders; cannot be balanced	replace bearing
resonance	1x, 2x, 3x, etc.	4.5	high balance sensitivity, high-amplitude vibration at order of operating speed	change structural natural frequency
looseness	1x plus large number of orders, 1/2x may occur	4.13	high 1x with lower-level orders, large 1/2 order, low axial vibration	shim and tighten bolts to obtain rigidity
eccentricity	1x	4.14	high 1x	machine journal for concentricity
coupling lockup	1x, 2x, 3x, etc.	4.12	1x with high 2x similar to misalignment; start and stops may yield different vibration patterns	replace coupling or remove sludge
thermal variability	1x		1x has varying phase angle and amplitude with load	compromise balance
distortion	1x and orders		1x from preload of bearings, 2x line frequency, air gap on motor	relieve soft foot
impact-excited natural frequency	Nx	4.5	excitation of natural frequency at multiple of operating speed	remove source of pulses or impacts
sheave eccentricity	1x		spectral content depends on pitch line profile. horizontal/vertical in phase	replace sheave

Rolling Element Bearing Defects

Fault	Spectral Frequency	Time Waveform/ Spectrum Shape	Comment	Page #
outer race defect	BPFO and multiples	multiples of BPFO with sidebands of RPS, FTF, or BSF	remaining life sensitive to speed and load	4.4
inner race defect	BPFI and multiples	multiples of BPFI with sidebands of RPS, FTF, or BSF	remaining life sensitive to speed and load	4.18
cage defect	FTF and multiples	FTF and multiples or sidebands on BPFO or BPFI	worst defect; failure imminent	4.19
ball defect	BSF or FTF and multiples	may be sidebands of BPFO or BPFI	limited time to failure	4.19
excessive internal clearance	natural frequencies	natural frequencies modulated by RPS	excessive clearance from wear	4.19

Gearbox Faults

Fault	Frequency	Page #	Spectrum Time Waveform
eccentric gears	gear mesh	4.23	gear mesh with sidebands at frequency of eccentric gear
gear-mesh wear	gear mesh and harmonics	4.3	gear mesh with sidebands at frequency of worn, scored, or pitted gear(s); sometimes 1/2, 1/3, 1/4 harmonics of gear mesh
improper backlash of end float	gear mesh and harmonics	4.5	gear mesh with orders and sidebands at frequency of pinion or gear
broken, cracked, or chipped gear teeth	natural frequencies	4.23	pulses in time waveform; natural frequencies in spectrum
gearbox distortion	gear mesh and/or natural frequencies	4.24	gear mesh and orders in spectrum; varying gear-mesh amplitude in time waveform – shaft frequency plus low-amplitude orders
common factor wear	fractional gear mesh and/or natural frequencies	4.3	fractional gear mesh frequencies dependent on the common factor(s)

Pumps and Piping

Fault	Frequency	Time Waveform/ Spectrum	Correction
structural resonance/piping and vertical pump	1x or vane pass	vibration focused at RPM or vane pass	change natural frequency/speed
acoustical piping resonance	vane pass	vibration focused at vane pass	change piping length or install damper
impeller eccentricity	1x or vane pass	vibration focused at 1x RPM	install new impellers
excessive impeller-diffuser clearance	1x, vane pass	vane-passing frequencies and multiples	change clearances
recirculation	random noise	random noise on data	change operation to best efficiency point (BEP)
cavitation	random noise, vane pass	random noise on data	change operation to BEP
excessive wear ring clearance	1x	strong 1x	may alter critical speed, damping, and stability

Fans

Fault	Frequency	Time Waveform/ Spectrum	Correction
structural resonance	1x, blade pass (BP)	operating speed/ BP	change isolators
structural cracks in impeller wheel	1x	nonrepeatable phase on balancing	repair wheel
loose wheel on shaft	1x and multiples	erratic phase	repair shaft
aerodynamic problems	1x, BP	random noise and vane pass	change damper position or ducting
pedestal stiffness, asymmetry	1x	large difference in horizontal-to-vertical amplitude	stiffen pedestal
belts and pulleys	1x, 2x, belt frequency	belt pulses, 1x vertical-horizontal in phase	replace pulley or belt
acoustical duct resonance	1x, vane pass	vibration focused at one fan speed	change ducting or fan speed

Motor Malfunctions — Electrical Faults

Refer to 4.25

Fault	Frequency	Page #	Spectrum; Time Waveform/Orbit	Correction/Comment
air-gap variation	120 Hz	4.28	120 Hz plus sidebands, beating 2x with 120 Hz	center armature relieving distortion on frame; eliminate excessive bearing clearance and/or any other condition that causes rotor to be off center with stator
broken rotor bars	1x	4.28	1x and sidebands equal to (number of poles x slip frequency)	replace loose or broken rotor bars
eccentric rotor	1x	4.29	1x, 2x/120-Hz beats possible	may cause air-gap variation
stator flexibility	120 Hz		2x/120-Hz beats	stiffen stator structure
off magnetic center	1x, 2x, 2x	4.29	impacting in axial direction	remove source of axial constraint-bearing thrust, coupling
stator shorts	120 Hz and harmonics	4.30	120 Hz and harmonics	replace stator

Overall Vibration Guidelines for Condition Evaluation Using Pedestal Measurements

Condition	Limits	
	rms velocity	peak velocity
acceptance of new or repaired equipment	<0.08	<0.16
unrestricted operation — normal	<0.12	<0.24
surveillance	0.12-0.28	0.24-0.7
unsuitable for operation	>0.28	>0.7

Service factors may be necessary for some special equipment, depending on design, speed, and process. Multiply service factor by measurement and check against Table. Limits are not valid for faults in gears and bearings.

Evaluation of Rotor/Bearing Vibration

Maintenance	Allowable R/C	
	3,600 RPM	**10,000 RPM**
normal	0.3	0.2
surveillance	0.3-0.5	0.2-0.4
shut down at next convenient time	0.5	0.4
shut down immediately	0.7	0.6

Table relates journal bearing clearance, rotor speed, and relative vibration to recommended maintenance actions. The ratio of the measured relative vibration (mils peak to peak) to the diametral clearance (mils peak to peak) of the bearing (R/C) is calculated and identified in the Table according to machine speed. Diametral clearance is the difference between the bearing diameter and the journal diameter.